"Show Me What You Know"

Exploring Student Representations Across STEM Disciplines

EDITED BY

Bárbara M. Brizuela

Brian E. Gravel

Foreword by Gerald A. Goldin

Teachers College, Columbia University
New York and London

Published by Teachers College Press, 1234 Amsterdam Avenue, New York, NY 10027

Library of Congress Cataloging-in-Publication Data can be obtained at www.loc.gov.

ISBN 978-0-8077-5409-2 (paper)

Printed on acid-free paper
Manufactured in the United States of America

20   19   18   17   16   15   14   13   8   7   6   5   4   3   2   1

Contents

PART III: HIGHLIGHTING

PART IV: REPRESENTATIONS AS SCAFFOLDS AND SUPPORTS

Foreword

The study of *representation*—fundamental to understanding learning, teaching, and problem solving—has long been an important theme in mathematics and science education. Having engaged in this study over several decades in the context of mathematical learning and problem solving, I have learned how complex the endeavor is—not least because the notion of representation has so many different aspects and interpretations. As the reader approaches the present research collection, spanning a number of the Science, Technology, Engineering, and Mathematics (STEM) disciplines, it may be helpful to keep in mind some of those distinct interpretations and the problematic but intriguing questions associated with them.

Research on representation during the 1980s and 1990s (e.g., Goldin & Janvier, 1998; Janvier, 1987) helped set the stage for the National Council of Teachers of Mathematics (NCTM) in the United States to include "Representations" as a major strand in its *Principles and Standards for School Mathematics* (NCTM, 2000), and to devote its 2001 Yearbook to the subject (Cuoco & Curcio, 2001). The NCTM offers a concise statement that points to different, but related meanings of representation:

> The term *representation* refers both to process and to product—in other words, to the act of capturing a mathematical concept or relationships in some form and to the form itself. Moreover, the term applies to processes and products that are observable externally as well as those that occur "internally," in the minds of people doing mathematics. (NCTM, 2000, p. 67)

Thus, "representation" sometimes refers to a tangible *production* or construct—such as a number line, a graph, a drawing, an arrangement of objects, a physical model, or a configuration on a computer screen. (Here one also may include written or spoken words, formulas or equations produced, and physical gestures as representations.) When the intent is to focus on a particular, concrete example of such a production, *without* tacitly supposing any assignment of meaning, semantic content, or correspondence with internal or "mental" states, the term *inscription* is often used. Thus, the term *representation* or *external representation* (i.e., external to the person who produced it) applies when one wants to *include* reference to some actual or potential representing relationship or meaningful interpretation.

In addition, "representation" can refer to the *process* of producing such an external representation, or it can refer to the *relationship* between the production and a mathematical concept that it may be said to "capture."

Alternatively, "representation" can refer to a *cognitive construct* or configuration (Palmer, 1978); in this context, one may refer to an "internal representation" (i.e., internal to an individual). The term *external representation* can be rendered somewhat more precisely in German as *Darstellung*, and *internal representation* as *Vorstellung* (von Glasersfeld, 1987). Moreover, representation can refer to the *cognitive process* of forming such an internal construct, or interpreting it in relation to other cognitive constructs.

Whether internal or external, representations may be based on *shared* conventions and interpretational schemes, or they may be *idiosyncratic* to a greater or lesser degree.

And representations rarely or never occur in isolation from other, related representations. On the contrary, they typically occur within structured *systems of representation*. In the case of conventional or shared representations, one can sometimes elucidate such representational structure in considerable detail, while in the case of idiosyncratic representations, structure may be more difficult to ascertain. *Ambiguity* and *context-dependence* are characteristic features of the interpretation of representations and structured systems of representation.

Internal or cognitive representational systems commonly considered include natural language, visual imagery, kinesthetic representation, learned formal notational systems, planning and executive control, and (sometimes) affective representation (e.g., Duval, 2006; Goldin, 1998, 2008; Goldin & Kaput, 1996; Hitt, 2002; Skemp, 1982). From this standpoint, the phenomenological primitives of diSessa (1983) may be regarded as internal imagistic and kinesthetic representations, nonstandard but not entirely idiosyncratic, *inferred* from external representations (drawings, gestures, descriptions, and so on) produced by students.

Each internal representational system develops, of course, in interaction with external systems; the latter include natural spoken and written language, notational systems, systems of pictorial and graphical representation, and so forth. And all of this takes place in social and cultural environments that influence and shape every aspect of representation (e.g., Anderson, Scheuer, Pérez Echeverría, & Teubal, 2009).

External representations are inscribed or embodied in media, and observation of how they are produced, manipulated, and transformed allows one to explore the representational fluency of learners and problem-solvers. As Lesh and Doerr (2003) observe,

> because different media emphasize (and deemphasize) different aspects of the systems they were intended to describe: (a) meanings associated with a given conceptual system tend to be distributed across a variety of representational media, (b) representational fluency underlies some of the most important abilities associated with what it means to understand a given conceptual system, and (c) solution processes [in problem solving] often involve shifting back and forth among a variety of relevant representations. (p. 12)

It should also be highlighted in this connection that with advancing technology, *action* representations or *dynamic* representations—those that change in response to actions according to built-in structures, or without the necessity of further agency—are playing an increasingly important role in STEM learning and teaching (e.g., Moreno-Armella, Hegedus, & Kaput, 2008).

A crucial question, then, may be stated in deceptively simple terms: "What do representations represent, and what is the nature of the representing relationship?" Different researchers have proposed and explored very different answers to this question, to which I cannot do justice in a short foreword. One point of view is that the distinction between "external" and "internal" representations is important, and that the representing relationship can be understood as bridging the external and the internal. A different view is that this distinction constitutes an unacceptable form of Cartesian "mind-body" dualism.

However, whenever we infer something about a student's cognition from his or her production of an external representation, or arrive at an inference about how the student has interpreted a representation, or study the uses of representations as forms of communication, we must, at least tacitly, consider what it means for the representation at hand to be "representing" something. Evidently, we cannot naïvely assume a close correspondence between features of an external representation produced by or presented to a student and features of the student's cognitive state. On the other hand, we must take there to be *some* relationship, and presumably it is one that involves meaning, symbolization, or semantic content—otherwise the notion of "representation" becomes superfluous.

This issue comes up a number of times in the present book, and the reader will find in its chapters some distinct ways of approaching it.

Of course, no one volume can address all the important aspects of representations. *"Show Me What You Know"* focuses mainly on external representations in the STEM disciplines, produced by or in some cases presented to learners. A key idea behind the book is that connecting representations across different disciplines has implications for representation within disciplines—and indeed, there is considerable promise to this suggestion. Mathematics education, for example, often relies on standard systems of representation: symbolic notation, number line and graphical representation, and so forth. In this volume, one finds a laudable openness to the exploration of idiosyncratic representations produced by individuals as they seek to understand or explain mathematical ideas or scientific phenomena. Indeed, this implements the editors' expressed philosophy of starting "where learners are." Then the representations produced by students, particularly when they are nonstandard, provide us with insights as to their conceptions that are easily overlooked when we restrict ourselves to drawing inferences through the lenses of standard representational systems.

The book is empirically rich, and not constrained to any one theoretical view. Its descriptions of ways in which representations in STEM disciplines are interpreted, and of new representations that are generated, provide opportunities for the reader to draw inferences about cognition and learning—students' conceptions and misconceptions, metarepresentational competencies, and so forth. The chapters move from an emphasis on the appropriation of (pre-existing) representations to processes of "making meaning" (of new as well as pre-existing representations). There is a focus on "highlighting" or "foregrounding"—how specific features of representations contribute to meaning-making and interpretation, and on how earlier, established representations or systems of representation can function as scaffolds for the construction of new ones.

Many of these ideas have been discussed theoretically in earlier literature. Here, the reader will find them richly and subtly applied in the context of empirical studies across different but related disciplines. We are provided not only with valuable source material for future theoretical development, but with profound encouragement for teachers and researchers to pay close attention to representations as they are generated and interpreted by students.

—Gerald A. Goldin,
Rutgers University, Graduate School of Education

REFERENCES

Anderson, C., Scheuer, N., Pérez Echeverría, M. P., & Teubal, E. V. (Eds.). (2009). *Representational systems and practices as learning tools.* Rotterdam, Netherlands: Sense Publishers.

Cuoco, A. A., & Curcio, F. R. (2001). *The roles of representation in school mathematics: NCTM 2001 Yearbook.* Reston, VA: National Council of Teachers of Mathematics.

diSessa, A. (1983). Phenomenology and the evolution of intuition. In D. Gentner & A. Stevens (Eds.), *Mental models* (pp. 15–33). Hillsdale, NJ: Lawrence Erlbaum Associates.

Duval, R. (2006). A cognitive analysis of problems of comprehension in a learning of mathematics. *Educational Studies in Mathematics, 61*, 103–131.

Goldin, G. A. (1998). Representational systems, learning, and problem solving in mathematics. *Journal of Mathematical Behavior, 17*(2), 137–165.

Goldin, G. A. (2008). Perspectives on representation in mathematical learning and problem solving. In L. D. English (Ed.), *Handbook of international research in mathematics education* (2nd ed.) (pp. 176–201). London, UK: Routledge.

Goldin, G. A., & Janvier, C. (Eds.). (1998). Representations and the psychology of mathematics education, parts I and II (Special Issues). *Journal of Mathematical Behavior, 17*(1–2).

Goldin, G. A., & Kaput, J. J. (1996). A joint perspective on the idea of representation in learning and doing mathematics. In L. Steffe, P. Nesher, P. Cobb, G. A. Goldin, & B.

Greer (Eds.), *Theories of mathematical learning* (pp. 397–430). Hillsdale, NJ: Lawrence Erlbaum Associates.

Hitt, F. (Ed.) (2002). *Representations and mathematics visualization*. México: Departamento de Matemática Educativa del Cinvestav—IPN.

Janvier, C. (Ed.). (1987). *Problems of representation in the teaching and learning of mathematics*. Hillsdale, NJ: Lawrence Erlbaum Associates.

Lesh, R. A., & Doerr, H. M. (2003). Foundations of a models and modeling perspective on mathematics teaching, learning, and problem solving. In R. A. Lesh & H. M. Doerr (Eds.), *Beyond constructivisim: Models and modeling perspectives on mathematics problem solving, learning, and teaching* (pp. 3–34). Mahwah, NJ: Lawrence Erlbaum Associates.

Moreno-Armella, L., Hegedus, S. J., & Kaput, J. J. (2008). From static to dynamic mathematics: Historical and representational perspectives. *Educational Studies in Mathematics, 68*(2), 99–111.

National Council of Teachers of Mathematics. (2000). *Principles and standards for school mathematics*. Reston, VA: Author.

Palmer, S. E. (1978). Fundamental aspects of cognitive representation. In E. Rosch & B. Lloyd (Eds.), *Cognition and categorization* (pp. 259–303). Hillsdale, NJ: Lawrence Erlbaum Associates.

Skemp, R. R. (Ed.) (1982). Understanding the symbolism of mathematics. Special issue, *Visible Language, 26*(3).

von Glasersfeld, E. (1987). Preliminaries to any theory of representation. In C. Janvier (Ed.), *Problems of representation in the teaching and learning of mathematics* (pp. 215–226). Hillsdale, NJ: Lawrence Erlbaum Associates.

Introduction

Bárbara M. Brizuela and Brian E. Gravel

Representations refer to both products and processes that we make or engage in as we attempt to capture, understand, and translate an idea, an event, or a phenomenon that we experience. For instance, a friend tries to explain to us the location of her new apartment—by remembering the neighborhood and the landmarks we know, we create a representation for ourselves of just where her new apartment is located. We might do this mentally, we might make a diagram on paper, we might use a map to confirm with our friend that we are thinking of the same location, or we might even engage in a combination of these activities. Representations are ubiquitous in our daily lives—we are constantly representing in some way or another. They also vary along several dimensions—they can be idiosyncratic and highly personal, or they may be conventional; they may have a physical nature or not; they may refer to a product or a process; they may reflect an idea or may have been used to construct that idea. Just as with representations in everyday life, we show in this book that representations (and their many dimensions) are ubiquitous to science, technology, engineering, and mathematics—the STEM disciplines.

The goal of this book is to showcase research on representations across a range of STEM disciplines and ages—from children as young as 2 years of age to professional mathematicians. We adopt the general perspective in this book that it is fundamental to "start where learners are." To do so, it is important for us to determine just what learners know and understand. Obviously learners include students, as the title of this book suggests; a discussion of learners more generally underscores the centrality of the themes in this book. This volume highlights the importance of paying close attention to learners' interpretations and productions of different representations; they are a source of evidence for what learners understand, and another way for learners to "show us what they know." Doing so provides educators and researchers with powerful insights; it also allows us to design interventions that more carefully consider students' own ways of doing and knowing. This perspective is theoretically framed by constructivism—as first espoused by Piaget (1926/1976,

1

1936/1952) in his emphasis on describing what children can do and explaining how children get from one way of knowing to another way of knowing, as well as in other more recent work related to constructivism and representations (e.g., Confrey, 1991; diSessa, Hammer, Sherin, & Kolpakowski, 1991; Kaput, 1991).

In addition to these assumptions, we also emphasize in this book the fundamental role played by social interactions and contexts, framed by the work of Vygotsky (1978), and Cole and Bruner (1971). It is clear that we build our own personal and individual representations and ways of symbolizing. Some representations and practices remain personal/private, while others make their way to the social arena. This book illustrates how in the social arena, we engage with conventions, audiences, communicative goals and intentions, and the construction of shared meanings mediated by the representations through which we interact. In the STEM fields, representations play a central role in the fundamental practices of these disciplines; at the core, the sharing and debating of ideas happens through representations—those we produce and those we interpret—and, thus, have a profound impact on our thinking in social contexts.

There are two claims that are woven throughout the work presented in this volume. The first claim is that making connections about the appropriation and construction of representations *across* the STEM disciplines is necessary, helpful, and productive for our understanding of the appropriation and construction of representations *within* each one of these disciplines. The different STEM disciplines can provide one another with insights and perspectives regarding representations as both products and processes. The specificities of one discipline can help illuminate aspects of other disciplines that we may not have previously noticed.

Navigating our way through the chapters in this volume led us to identify four themes—not meant to be limiting, but rather broadening and generative—that have emerged as organizing and emphasizing the work of our colleagues in different ways: Appropriation of Representations, Making Meaning, Highlighting, and Scaffolds and Supports. In a way, as editors of this volume, we have engaged in our own process of representing the content of our colleagues' work by proposing one possible organization. As in Julio Cortázar's epic novel *Hopscotch* (1963/1966)—in which alternative orderings of the chapters proposed by Cortázar himself or by the reader lead to different novels and readings—alternative readings, sequencing, and ways of organizing the chapters might have led to different meanings being constructed of the "same" content. The second claim that we put forth in this volume is that these four themes or lenses help us to critically examine the multitude of ways in which we interact with representations in the STEM disciplines. They also provide structure for engaging with the works presented in this book and beyond this volume as well.

Each of the four themes is captured in a part of the book; each part contains three chapters and each, in turn, ends with a reflection on the issues each chapter raises. The first theme, *Appropriation of Representations*, refers to how learners immersed in a world inundated with representational forms must come to understand and appropriate ways of externalizing. From our perspective, invention and appropriation are crucial aspects to how learners come to understand ways of expressing their ideas. Following Kaput (1991), one way to think about appropriation processes is to note that the act of representing helps us to organize and refine our ideas, and the artifact we generate becomes a vehicle for thought. An example of *Appropriation* taken from this first section is that provided by Anthony's drawing shown in Figure 3.14, in Kahn's chapter. In this drawing, Anthony is showing that he has appropriated the implicit rule underlying some representational systems, in which more of something is represented by more of something else. In his production in Figure 3.14, Anthony uses the lines behind a car to depict speed—and "more speed" (or "faster") is represented using more lines.

The second theme is that of *Making Meaning*. By *Making Meaning*, we refer to the complex processes that learners face in becoming fluent users of representations. Central to these processes is learners' making meaning, interpreting, and expressing their understandings in increasingly logical and systematic ways. An example of *Making Meaning* from this section of the book is that of the interactions that occur among the group of middle school boys in Wright's chapter as they try to make sense of the representation created by Earl and shown in Figure 6.1. Earl proposed a new way of using waves to represent sound, one that ran counter to the group's previously agreed-upon use and meaning. The crux of the interaction detailed in Excerpt 1 in Wright's chapter—arguing about the use of a specific symbol—illustrates the fleeting and complex process of making meaning of representations.

The third theme is *Highlighting*. Goodwin (1994) proposed that representations highlight ways in which we "see" particular phenomena. In other words, the representations we produce reflect the aspects, characteristics, boundaries, and processes we find most pertinent to the inquiry at hand. An example of *Highlighting* is that proposed by Nemirovsky and Smith in their chapter. Their interviewee, Joseph, is comparing different spin combinations in quantum physics. In his explanations and re-representation of the issue, Joseph creates a tetrahedron to highlight specific aspects of the spin combination that were previously implicit to both him and the interviewer. This chapter does not specifically showcase students' construction and use of representations; however, it provides a detailed and sophisticated example of the highlighting theme, which we believe to be useful in considering students' engagements with representations.

The fourth and last theme, and section of this book, is *Representations as Scaffolds and Supports*. *Scaffolds and Supports* builds from a Vygotskyian (1978) perspective on tool use, specifically the idea of representations as mediators of one's activities. Representations can be scaffolds and supports for engaging with new ideas, for contemplating familiar ideas, and for reexamining how we communicate our thinking in different forms. An example of *Scaffolds and Supports* from this section is that proposed by Schliemann, Carraher, and Caddle. In their chapter, they illustrate how children can use parallel number lines as a scaffold and support for understanding the representation of co-varying quantities in the Cartesian space.

Finally, Schwartz's chapter invites us to extend our thinking to the implications of representations and models in the specific area of science education.

While we have just presented the distinctions between the four themes and lenses as stark and neat, they are far from that. Each lens provides us with different ways of thinking about representations. The distinctions, however, provide us with an opportunity to bring to the foreground specific themes and lenses. In turn, the distinctions background other issues that, while not the object of discussion, certainly remain.

We hope that while reading this volume you will begin to ask yourself, "How do I interact with representations (others' and my own) in my daily life?" "What role do they play?" "How do I make meaning of them?"

ACKNOWLEDGMENTS

We are deeply thankful to the following reviewers for their suggestions and comments to drafts of the chapters in this book: Andrés Acher, Alfredo Bautista, Cindy Ballenger, Maria Blanton, Christine Cunningham, Liz Gunderson, Winnie Ko, Andee Rubin, Senay Purzer, Bruce Sherin, Roger Tobin, David Uttal, and Marianne Wiser. Additionally, we'd like to thank Mary Caddle and Pearl Emmons.

REFERENCES

Cole, M., & Bruner, J. S. (1971). Cultural differences and inferences about psychological process. *American Psychologist, 26*(10), 867–876.

Confrey, J. (1991). Learning to listen: A student's understanding of powers of ten. In E. von Glasersfeld (Ed.), *Radical constructivism in mathematics education* (pp. 111–138). Dordrecht, Netherlands: Kluwer Academic Press.

Cortázar, J. (1966). [1963]. *Hopscotch* (G. Rabassa, Trans.). New York: Pantheon Books.

diSessa, A. A., Hammer, D., Sherin, B., & Kolpakowski, T. (1991). Inventing graphing: Metarepresentational expertise in children. *Journal of Mathematical Behavior, 10*, 117–160.

Goodwin, C. (1994). Professional vision. *American Anthropologist, 96*(3), 606–633.

Kaput, J. (1991). Notations and representations as mediators of constructive processes. In E. von Glasersfeld (Ed.), *Radical constructivism in mathematics education* (pp. 53–74). Dordrecht, Netherlands: Kluwer.

Piaget, J. (1952). [1936]. *The origins of intelligence in children* (M. Cook, Trans.). New York: International Universities Press.

Piaget, J. (1976). [1926]. *The child's conception of the world* (J. & A. Tomlinson, Trans.). Totowa, NJ: Littlefield, Adams.

Vygotsky, L. S. (1978). *Mind in society: The development of higher psychological processes*. Cambridge, MA: Harvard University Press.

Part I

APPROPRIATION OF
REPRESENTATIONS

CHAPTER 1

Symbolic Use of Quantitative Representations in Young Children

Eduardo Martí, Nora Scheuer,

and Montserrat de la Cruz

In this chapter we present a study that addresses the following question: At what age and through what cognitive processes do children show evidence of symbolic uses of graphic representations of discrete quantities?

FIRST APPROACHES TO GRAPHICAL REPRESENTATIONS OF QUANTITIES

In order to frame the study, we first briefly review related previous research. Children establish contact with quantities and numbers very early, usually before they start school (Gelman & Gallistel, 1978; Tolchinsky, 2003). From the age of 2 or 3, children in highly literate environments perceive and show interest in the numerals they encounter in different contexts within their environment.[1] Children manage to identify these forms, attribute various functions to them, and distinguish them from other notations, such as writing and drawing (Sinclair, 1988; Tolchinsky, 2003). However, it is not clear whether children younger than 4 years of age understand the symbolic nature of these graphic forms, that is, whether they understand that according to their formal properties, numerals carry a precise meaning related to quantity (the number of simple or composed units), order (the place in a series where something is found), or identification (what a particular object is) (see Sinclair & Sinclair, 1984). Before the age of 4, the graphic forms that children produce in order to represent quantities might not have a completely symbolic nature yet. According to Bialystok and Codd (1996), one of the conditions for these notations to achieve a symbolic nature is that the notation must be given a stable numerical meaning. If, for example, the child produces a graphic form to represent that a box has three elements, moments later this same graphic form must serve to indicate that there are three elements in the box. It appears that this stability is closely related to the production of conventional numerals. From 4 and 5 years of age, children produce some conventional numerical notations, depending on factors such as their familiarity with these cultural forms, their goals in relation to the notational activity, the magnitude of the numbers involved, and the ways in which the quantified objects are conveyed (verbal, physical, graphic) (Dockrell & Teubal, 2007). Children's ways of noting quantities may even change as they perform a given task, indicating that their understanding of the notational and communicative demands and restrictions may evolve in very short time spans (Scheuer & Germano, 2005).

Before the age of 4 or 5, one of the most frequent ways in which children represent discrete quantities is based on the iteration of graphic forms. This procedure is regulated according to one-to-one correspondence between graphic forms and objects. That is, a same graphic form is repeated (or iterated) as many times as the number of objects to be represented on paper (e.g., /// for three sweets). In this way, each object is represented by means of one graphic form (García-Milà, Teberosky, & Martí, 2000; Hughes, 1986; Klein, Teubal, & Ninio, 2009; Sinclair, 1988). According to this representational approach, if a single object is represented with a single discrete symbol, two objects are represented with two symbols, and so on, then the absence of objects corresponds to the absence of symbols. These kinds of notations, which allow for the meaning of the notation to be easily reconstructed after the time of production (Martí, 2003), are invoked throughout life in various situations, especially to account for collections of objects that keep increasing. Some authors have indicated that this kind of representation was important in the historical development of numerical systems (Ifrah, 1994; Olson, 1994). The questions that remain to be answered are when and how children attribute a symbolic function to these iterative forms of representation.

In addition to the number domain, the works of DeLoache (1991, 1995) inquire into the origin of children's understanding of the symbolic nature of a range of figurative graphic representations (maps, photographs, drawings). Determining whether children consider a graphic representation to be a symbol involves analyzing whether they are capable of using the spatial information provided by the representation (photo or drawing) to act in an appropriate manner in another space (for example, to find the precise place where a doll is hidden by using the graphic representation or photograph standing for the doll). It is important to note that for this to occur, according to DeLoache, children must understand that graphic representations have a double nature (what DeLoache calls "dual representation"): They are physical objects with a given form, color, texture, and position in space, yet at the same time they represent properties that refer to, and therefore can be transferred to, other situations. Moreover, children must understand that these very specific physical and spatial properties (e.g., close/far, right/left, in front of/behind relationships) allow them to move from the representational space to the space that is being represented, or the target space. According to studies in the spatial domain, this capacity to use graphic representations as symbols becomes consolidated between 2 and 3 years of age (DeLoache, Miller, & Pierroutsakos, 1998; del Rosario Maita & Peralta, 2009; Peralta & DeLoache, 2004).

DeLoache's studies are mainly concerned with figurative symbols (those that establish a relatively direct correspondence between the properties of the symbol and those of the referent). It should be expected that with graphic representations of a more abstract nature, such as quantitative representations, the symbolic relationship between notation and meaning will develop more slowly. Nieder and Dehaene (2009) portray this abstract nature of number very clearly: "Number, however, is an abstract category devoid of specific sensory features; two cats and two calls have nothing in common, except that the size of their sets is two" (p. 192).

To summarize, precedent research has shown that until the age of 4, children have difficulties in understanding the symbolic value of conventional notations of number. However, as far as we know, toddlers' understanding of the symbolic value of iterative notations of number has not been systematically addressed in the literature.

NUMERICAL SKILLS PRIOR TO THE USE OF NUMERALS

Well before children are able to produce utterances, they reveal an awareness of a wide range of quantitative variations, regularities, and correspondences. Prior studies have shown that infants and even newborns discriminate among pairs of visual-spatial arrays, as long as they do not exceed the quantity of three (i.e., two versus three dots, but not four versus six dots) (Antell & Keating, 1983; Starkey & Cooper, 1980). When babies are close to their first birthday, they demonstrate this ability when dealing with arrays of figurative drawings (Strauss & Curtis, 1981) and arrays of one, two, and three physically present objects (Feigenson, Carey, & Hauser, 2002). Many studies have broadened the picture of infants' quantitative discrimination abilities to include pairs of small arrays of figures in motion (Van Loosbroek & Smitsman, 1990), auditory stimuli (Bijeljac-Babic, Bertoncini, & Mehler, 1991), or actions (Wynn, 1996), and pairs of larger quantities or auditory sequences differing by a ratio of 2 or 1.5 (e.g., 8 versus 16 and 8 versus 12) (Lipton & Spelke, 2003). Furthermore, babies establish correspondences between displays presented in two different sensory modalities (Koyabashi, Hiraki, & Hasegawa, 2005; Starkey, Spelke, & Gelman, 1990) and seem to anticipate the result of adding or subtracting one object from a small array correctly (Wynn, 1992). All of these cases deal with reactions, anticipations, or comparisons involving quantitative primary information, which is not conveyed by conventional quantitative symbols.

With the advent of language, children begin to manage verbal representations of numbers. As they begin to speak, children use numerical terms, especially in relation to collections of one, two, or three elements, which in some cases are their own hands or fingers (Fuson, 1988; Mix, Sandhofer, & Baroody, 2005; Scheuer & Sinclair, 2009; Sinclair, 2005; Wynn, 1990, 1992). One of the functions children use these numerical terms for is to measure a small collection or to compare it quantitatively to another one, something that young children can and often do without counting. During the preschool years, this "virtually instantaneous apprehension of number" (Fischer, 1992, p. 191), or subitizing, involves quantities up to three. Several studies have shown that 2- to 4-year-old children can match two small sets of objects based on the set size (Mix, 2008); children can also match sets that are sequential (actions) and also sets across modalities (for example, objects to actions). These last cases are more challenging than the ability to match objects to objects (Mix, 1999; Mix, Huttenlocher, & Levine, 1996).

Approximately by the age of 2 or 3, children begin to display counting behaviors before they use this procedure to determine or construct specific quantities. The studies of Gelman and Gallistel (1978) revealed the principles that govern counting among children at these ages. However, knowing how to count collections formed by any type of objects, by assigning to each of them one (and only one) specific word taken from an ordered and stable series (one, two, and so on), does not guarantee that children infer that the last assigned word expresses the total quantity. As Sophian (1996) has shown, it is mostly starting at the age of 4 that children come to use the result of counting to know whether one collection is more numerous than another, or to form another collection with the same number of elements.

The review of the above studies highlights one pending issue: determining at what age and through what cognitive processes children are capable of using graphic representations of numbers in a symbolic form. We know that between 2 and 3 years of age, children are capable of symbolically using figurative graphic representations to identify and locate objects in a target space, which indicates a pragmatic understanding of the symbol-referent relationship. Yet, as far as we know, no study has explored this question taking into consideration graphic representations of number. This is the central objective of the present study. In the next section, we explain the assumptions upon which the adopted design is based and outline the specific aims we pursue.

Seeking to Capture Children's First Uses of Graphic Representations of Quantity

One of the earliest ways of representing quantity is through iterative notations (in which one-to-one correspondence is established between the elements in the notation and the elements in the collection). Thus, we believe that examining children's first uses of graphic representations of quantity must rely on this type of notation, the interpretation of which appears to be relatively simple and direct compared with the interpretation of notations derived from the decimal positional system that bears a mostly arbitrary symbolic relation with the represented numbers (see Martí, 2003).

Because babies are capable of establishing distinctions among displays of one, two, and three objects (e.g., dots, sounds, or actions, as reported above) from the first months of life, we restricted our inquiry of the first symbolic uses of graphic representations of quantity to this one-to-three interval. We also focused our inquiry on the first uses of graphic representations of quantity among children between 3 and 4 years of age, keeping in mind that at the age of 3 years, children are already capable of: (1) using spatial information symbolized through figurative representations, and (2) describing the quantification of small amounts, even if they do so through methods other than counting as subitizing.

In terms of the first point, as shown in the spatial studies by DeLoache and her colleagues, to show that a child is capable of dealing with a graphic representation as a symbol, it is necessary for the information conveyed by the graphic representation to be abstracted and applied to another situation. The studies performed until now in the domain of number have explored skills related to collections of objects (counting them, comparing two collections, constructing an equivalent numerical collection, and so on), but they have not evaluated if and how young children are capable of using the numerical information expressed by graphic representations. An example of this kind of task would be to present the child with iterative graphic marks on a card and to ask him or her to fetch as many dolls from a basket as marks appear on the card.

If a graphic representation is taken as a symbol, it is possible to establish two types of relationships: (1) start from the graphic representation and apply the numerical information it conveys to a situation; and (2) start from the situation to choose the graphic representation that represents it. Effective performance in both directions of the relationship (from the symbol to the referent and from the referent to the symbol) is important, but we do not know if the difficulty they pose to children is identical.

Based on these considerations, through this study we sought to:

1. Determine at what age and through what cognitive processes children are capable of symbolically using an iterative representation of quantity (one, two, three, and an empty set).
2. Compare the process through which children interpret a graphic representation (from the representation to the referent) with the process that allows them to go from the situation to the selection of the appropriate representation (from the referent to the representation).

Inviting Children to Use Graphic Representations of Quantity

Twenty-four children with ages ranging from 2 years, 11 months (2y,11m) to 4 years were encouraged to participate in a simple game, consisting of making a 10-centimeter-high plastic toy horse advance step by step along a rubber-made path. The path measured in centimeters (cm) was 10 cm x 300 cm, on which 20 steps were marked, each containing a yellow oval and separated from the adjacent ones by black vertical lines. The game was based on a very simple rule, establishing that the quantitative information provided by a red, wooden die indicated the number of steps the horse should be moved forward along the path. The faces of the die (measuring 4 cm on each side) presented one, two, or three dots, and one face showed no dots.

Children's participation in the game took place within short individual interviews in a familiar environment. Children were distributed in three age ranges: eight children with ages between 2y,11m and 3y,2m (age 1); nine children aged between 3y,4m and 3y,7m (age 2); and seven children aged between 3y,9m and 4y (age 3). All of the children attended public or private education centers in San Carlos de Bariloche, Argentina.

Since young children may respond in very different ways according to very precise characteristics of the tasks they are asked to solve, we will carefully go through the procedure used in the interviews.

Demonstration of the Game. The interviewer presented the child with the die and asked: "What is this? What does it have on its faces?" Then she placed the die with the face containing two dots upward and moved the horse two steps, without saying anything. The demonstration was repeated for one and three dots.

Task 1. From graphic representation to referent: Request to advance the horse. The interviewer explained: "Now we begin the game. You will move the horse and I'll place the die [on the game board]. Pay attention; to move the horse, you have to look at the dots on the die. Now I'm placing the die [on the game board]."

Once each face was presented, the interviewer stated: "Look carefully at what the die shows here on top, and move the little horse."

The interviewer placed the die successively with the following faces upward: two, one, three, empty face, one, empty face, three, two, three, two, empty face, one. A fence was included in the latter part of the game.

Task 2. From referent to graphic representation: Request to place the die. In between the previous sequence of requests (after every four throws of the die), the interviewer placed the fence some steps ahead of the horse. She gave the die to the child, telling him or her, "Look, the horse is close to the fence. Look carefully to see how much is left [to reach the fence]. Now I'm going to ask you: Where must the die be for the horse to arrive at the fence? What face must the die show for the horse to arrive at the fence? Place the die [on the board game], please."

Throughout this task, children were faced with a situation where two, one, and three free steps respectively separated the horse from the fence.

Children's Ways of Participating in the Game

To analyze children's responses, we read the complete interview transcripts and observed the interview videos with the goal of delineating mutually excluding response categories for each one of the tasks, according to a grounded methodology. We distinguished between children's responses to the request to advance the horse from those to the request to place the die (see Table 1.1). Thus, two macrounits of analysis were considered: (1) the predominant behavior of the child for each face of the die in the course of the whole task, that is, all the requests to move the horse; and (2) the child's behavior when requested to place the die on the board game him- or herself (let us recall that there were three requests to place the die, to indicate the displacement across two, one, and three free spaces). This is because it is not possible to interpret the meaning of each isolated response. For example, if a child advanced the horse one step after each placement of the die, it would be absurd to consider that she was correctly solving the task for the face with one dot. Similarly, a behavior was coded as "global advancement" (see Table 1.1) based on the comparison of the advancements that the child made of the horse for the faces with more and less dots, but this cannot be established on the basis of the response to a particular presentation. Categories are explained in detail in Table 1.1.

In order to analyze the relationships among children's behavior upon the request to advance the horse (from graphic representation to referent) and to place the die (from referent to graphic representation) according to their age range (age 1, age 2, and age 3), we applied a multivariate analysis technique of a descriptive nature (Analysis of Multiple Correspondences; see Crivisqui, 1993). This analysis allowed us to identify "groups" formed by children who responded in similar ways to the various requests, as well as to assess if such groups were somehow related to children's age range.

On the basis of statistical criteria, we distinguished three groups formed by children and response categories obtaining a contribution to the formation of the axes that exceeded the average contribution (that is, not all the children "belong" in these groups; only those showing a distinct behavior).

The groups are the following:

Group 1. This group is formed by four of the youngest children, who either evidence lack of response to one or both tasks (Horse Advancement and

Table 1.1. Response Categories

From graphic representation to referent: Request to advance the horse	
Correct numerical advancement for one, two, and three	The child advances the horse the correct amount of steps for one, two, and three dots and does not move the horse for the face with no dots.
Correct numerical advancement for one and two, and *once* for three	The child advances the horse the correct amount of steps for: • all of the presentations of the face with one dot; • all of the presentations of the face with two dots (this includes responses that are two jumps in the same spot); and • only one of the three presentations of the face with three dots. On the other two occasions, the child advances or jumps two spaces instead of three. The child leaves the horse still for the presentations of the face without dots.
Correct numerical advancement for one and two	When shown the faces with one and two dots, the child moves the horse the correct amount of spaces. When shown the face with no dots, he leaves the horse still. When shown the face with three dots, he advances the horse an incorrect number of spaces.
Binary advancement	When shown faces with dots, the child moves the horse one space forward. When shown the face without dots, either she leaves the horse still, or places the die with a face containing dots up and moves the horse one space.
Advancement regulated by turns	After the interviewer places the die, the child advances the horse one space on the path independently from the information on the visible face of the die. He does so even for the face without dots.
Global advancement	After the interviewer places the die, the child moves the horse one or more spaces, up to five.
Unregulated advancement	The child advances the horse without taking into account the basic rule of the game, i.e., does not wait for the die to be placed or does not take the die into account.
No response	The child does not frame her activity within the advancement game.
Correct numerical	The child places the die with the face that shows how many steps separate the horse from the fence placed upward: • in the three sections of the task (when one, two, or three spaces remain); or • in the last two parts of the task (when one or three spaces remain), while he fails the first part of the task.

Table 1.1. Response Categories (continued)

From referent to graphic representation: Request to place the die	
Any face upward	In the three sections of the task, the child places the die with any of the faces with dots upward, without showing any concern for how many dots are on the face of the die.
The child does not use the die and/or she uses it as a landmark	The child moves the horse without placing the die in any position, or he places the die next to the fence instead of moving the horse. That is, he uses the die as a landmark instead of as an instruction.
Does not use or uses as landmark, and places any face upward	The child does not use the die or uses it as a landmark, and in at least one opportunity places the die on any face with dots.
No response	The child does not frame her activity within the game.

Placement of the Die), or make the horse advance along the path without considering the quantitative information provided by the die, do not use the die, or use it as a landmark under the request to indicate the steps the horse must move forward. This group is associated with the youngest age range.

Example: Matías, 3y,0m,9d

Task 1. Request to advance the horse

Matías moves the horse along the path without taking the die into account. He does not wait for the die to be placed or look at the dots on its faces at any moment.

Task 2. Request to place the die

Matías moves the horse up to the fence without using or touching the die.

Group 2. This group is formed by six children, four of whom have an intermediate age. These children evidence a binary advancement of the horse. For the request to place the die in order to move the horse forward, they either place it with any face upward, or show the following combination during the course of the game: not using the die to move the horse, using the die as a landmark, and placing it with any side up. The intermediate age is associated with this group.

Example: Marco, 3y,6m

Task 1. Request to advance the horse

In order to move the horse, Marco waits for the die to be placed and looks at the dots on its upper face. For all the placements showing one, two, or three dots, Marco attempts to quantify the amount of dots (he touches each one of them, says how many there are) and then moves the horse one step forward. For all the placements of the die showing the empty face upward, Marco comments on the absence of dots ("nada puntitos, no hay nada," which in English would read: "nothing dots, there is nothing"), turns the die around and places it with any of the faces with dots looking upward, and moves the horse one step forward.

Task 2. Request to place the die

For the first request, when there are two empty steps between the horse and the fence, Marco places the die on the path near the fence. For the other two requests (when there are one and three free steps), Marco places the die with any face with dots looking upward and then moves the horse forward step-by-step until it gets to the fence.

Group 3. This group is formed by five children in the three age ranges who evidence numerical advancement of the horse for one, two, and three dots as well as correct placement of the die to indicate the number of spaces from the horse until the fence.

Example: Antonella, 3y,5m

Task 1. Request to advance the horse

For the first placement of the die, Antonella remains still and does not move the horse. For all the following placements of the die, she advances the horse according to the number of dots on the die, after correctly identifying them verbally and/or through gestures. On one occasion she quantifies three dots as if there were two (there are two), and advances the horse accordingly, by two spaces.

Task 2. Request to place the die

When there are two spaces left, Antonella places the die on the side of the fence, without taking into account the amount shown on the face of the die that remains upward (which is empty), and moves the horse up to the fence. When there is one space left, she repeats this behavior (and places the face of the die with three dots upward).

> *Interviewer:* Look at how many spaces there are to the fence.
> *Antonella:* One.
> *Interviewer:* And how many dots are there on the die?
> *Antonella* [points at each dot]: One, two, three.
> *Interviewer:* And so what?
> *Antonella* [facially expresses sudden understanding]: One (dot) must come
>   out (on the die)!

She turns the die around until she finds the face with one dot, places it upward, and moves the horse one step forward. When there are three spaces left, she counts them, looks for the face with three dots, then places that face upward and moves the horse to the fence.

CONCLUSIONS

Our main objective was to develop a preliminary understanding of the age and cognitive processes through which children are capable of using graphic representations of number in a symbolic way. The results of this exploratory study indicate that 3- to 4-year-old children who are developing in literate environments manifest a wide range of approximations to the use of simple quantitative symbols. This range extends from not understanding the rules of the proposed game to the full mastery of the instructions of the game. The complete range of approximations can even be observed among the youngest children, those in the age range from 2y,11m of age to 3y,2m. The range of approximations is somewhat reduced among the older children; they all pay attention to the turns of the game, although with varying levels of integration of the quantitative information.

Taken together, these results indicate that some children begin to fathom the symbol-referent relationship for iterative notations of very small quantities at 3 years of age, whereas some of the older children have not yet managed to grasp the quantitative dimension of the game. These findings suggest that grasping the symbol-referent relationship is more challenging—or achieved later—in the case when discrete quantities are involved than in regard to spatial relations. In effect, the studies by DeLoache (1991, 1995) show that children between the ages of 2.5 and 3 years appropriately use basic spatial information represented in maps, photographs, or models. Our expectation that children's capacity to symbolically use quantitative notations—as explored in this study—would follow their ability to symbolically use figurative graphic representations to identify and locate objects in a target space—as shown by DeLoache's studies—appears to be supported.

Continuing with our proposed objectives, we found that the correct resolution of the tasks of advancement of the horse and of placement of the die are closely associated with one another. These results suggest that when children establish a symbolic relationship between the number of dots (representational space) and the number of spaces in the path (target space), this relationship works in both directions (from the dots to the spaces and from the spaces to the dots; that is, from symbol to referent and from referent to symbol). In future studies, it would be interesting to examine whether this symmetry is also found with other types of symbolic relationships in which such a direct correspondence between the characteristics of the symbol (number of dots) and the characteristics of the referent (number of spaces) is not present, as when Arabic numerals are presented.

The results suggest a remarkable discontinuity between babies' widely documented perceptive abilities to discriminate among collections of up to three elements and the grasp of the symbolic meaning of simple graphic representations of collections of the same magnitudes. In fact, only one of the three groups identified through the analysis was able to transfer the quantitative information conveyed by the die in order to use it to regulate the number of steps, or vice versa, correctly quantifying three dots (i.e., Group 3).

Although some of the younger children did not understand the basic rules of the game (placement of the die, looking specifically at the dots, and moving the horse forward), the majority of them participated in the game even though they may not have interpreted the amount of dots correctly. This indicates that beginning at 3 years of age, children can participate in a simple, regulated game. To the best of our knowledge, this activity has rarely been explored at these early ages, in which pretend play has been prioritized (Harris, 2000; Piaget, 1959; Vygotsky, 1978). The results appear to indicate that the understanding of simple rules of action that make up a game may appear very early. This ability surely has, as a precursor, the understanding of the conventional use of the objects, a process preceding the appearance of the first symbols (Rodríguez, 2006).

ACKNOWLEDGMENTS

The study was conducted with support from the Spanish Project EDU2010-21995-C02-02 (Ministerio de Ciencia e Innovación) and the following projects from Argentina: PICT 1607 (Agencia Nacional de Promoción Científica y Tecnológica), B139 (Universidad Nacional del Comahue), and PIP 1029 (CONICET).

NOTE

1. In this chapter, we use the term "*numeral*" to refer to graphically displayed digits or strings of digits, produced according to the Hindu-based decimal numerical notation system.

REFERENCES

Antell, S. E., & Keating, D. P. (1983). Perception of numerical invariance in neonates. *Child Development, 54*, 695–701.

Bialystok, E., & Codd, J. (1996). Developing representations of quantity. *Canadian Journal of Behavioural Science, 28*, 281–291.

Bijeljac-Babic, R., Bertoncini, J., & Mehler, J. (1991). How do four-day-old infants categorize multisyllabic utterances? *Developmental Psychology, 29*, 711–721.

Crivisqui, E. M. (1993). *Análisis factorial de correspondencias. Un instrumento de investigación en ciencias sociales.* Asunción, Paraguay: Universidad Católica de Asunción.

DeLoache, J. S. (1991). Symbolic functioning in very young children: Understanding of pictures and models. *Child Development, 62,* 736–752.

DeLoache, J. S. (1995). Early symbol understanding and use. *The Psychology of Learning and Motivation, 33,* 65–114.

DeLoache, J. S., Miller, K. F., & Pierroutsakos, S. L. (1998). Reasoning and problem solving. In D. Kuhn & R. Siegler (Eds.), *Handbook of child psychology, 5th ed., vol 2.: Cognition, perception, & language* (pp. 801–850). New York: Wiley.

Dockrell, J., & Teubal, E. (2007). Distinguishing numeracy from literacy: Evidence from children's early notations. In E. Teubal, J. Dockrell, & L. Tolchinsky (Eds.), *Notational knowledge: Historical and developmental perspectives* (pp. 113–134). Rotterdam, Netherlands: Sense Publishers.

Feigenson, L., Carey, S., & Hauser, M. (2002). The representations underlying infants' choice of more: Object files vs. analog magnitudes. *Psychological Science, 13*(2), 150–156.

Fischer, J. P. (1992). Subitizing: The discontinuity after three. In J. Bideaud, C. Meljac, & J. P. Fischer (Eds.), *Pathways to number: Children's developing numerical abilities* (pp. 191–207). Mahwah, NJ: Lawrence Erlbaum Associates.

Fuson, K. C. (1988). *Children's counting and concepts of number.* New York: Springer-Verlag.

García-Milà, M., Teberosky, A., & Martí, E. (2000). Anotar para resolver una tarea de localización y memoria. *Infancia y Aprendizaje, 90,* 51–70.

Gelman, R., & Gallistel, C. R. (1978). *The child's understanding of number.* Cambridge, MA: Harvard University Press.

Harris, P. L. (2000). *The work of the imagination.* Oxford, UK: Blackwell.

Hughes, M. (1986). *Children and number.* Cambridge, MA: Blackwell.

Ifrah, G. (1994). *Histoire universelle des chiffres.* Paris, France: Laffont.

Klein, E., Teubal, E., & Ninio, A. (2009). Young children's developing ability to produce notations in different domains—Drawing, writing, and numerical. In C. Andersen, N. Scheuer, M. del P. Pérez Echeverría, & E. Teubal (Eds.), *Representational systems and practices as learning tools* (pp. 39–58). Rotterdam, Netherlands: Sense Publishers.

Kobayashi, T., Hiraki, K., & Hasegawa, T. (2005). Auditory-visual crossmodal numerosity matching in infants and young children. *Japanese Journal of Psychonomic Science, 24,* 123–124.

Lipton, J. S., & Spelke, E. S. (2003). Origins of number sense: Large number discrimination in human infants. *Psychological Science, 14,* 396–401.

del Rosario Maita, M., & Peralta, O. (2009). El impacto de la instrucción en la comprensión temprana de un mapa como objeto simbólico. *Infancia y Aprendizaje, 33*(1), 47–62.

Martí, E. (2003). *Representar el mundo externamente. La adquisición infantil de los sistemas externos de representación.* Madrid, Spain: Antonio Machado.

Mix, K. S. (1999). Preschoolers' recognition of numerical equivalence: Sequential sets. *Journal of Experimental Child Psychology, 74*(4), 309–332.

Mix, K. S. (2008). Surface similarity and label knowledge impact early numerical comparisons. *British Journal of Developmental Psychology, 26,* 13–32.

Mix, K. S., Huttenlocher, J., & Levine, S. C. (1996). Do preschool children recognize auditory-visual numerical correspondences? *Child Development, 67,* 1592–1608.

Mix, K. S., Sandhofer, C. M., & Baroody, A. J. (2005). Number words and number concepts: The interplay of verbal and nonverbal processes in early quantitative development. In R. Kail (Ed.), *Advances in child development and behavior: Vol. 33* (pp. 305–346). New York: Academic Press.

Nieder, A., & Dehaene, S. (2009). Representation of number in the brain. *Annual Review in Neuroscience, 32,* 185–208.

Olson, D. R. (1994). *The world on paper: The conceptual and cognitive implications of writing and reading.* Cambridge, UK: Cambridge University Press.

Peralta, O., & DeLoache, J. S. (2004). La comprensión y el uso de fotografías como representaciones simbólicas por parte de niños pequeños. *Infancia y Aprendizaje, 27,* 3–14.

Piaget, J. (1959). *La formation du symbole chez l'enfant.* Neuchâtel, France: Delachaux et Niestlé.

Rodríguez, C. (2006). *Del ritmo al símolo. Los signos en el nacimiento de la inteligencia.* Barcelona, Spain: Horsori.

Scheuer, N., & Germano, A. (2005). Conocimientos matemáticos de niños de 4 a 7 años en entornos de alfabetización limitada. In M. Alvarado & B. Brizuela (Eds.), *Haciendo números. Las notaciones numéricas vistas desde la psicología, la didáctica y la historia* (pp. 109–145). México: Paidós.

Scheuer, N., & Sinclair, A. (2009). From one to two: Observing one child's early mathematical steps. In C. Andersen, N. Scheuer, M. del P. Pérez Echeverría, & E. Teubal (Eds.), *Representational systems and practices as learning tools* (pp. 19–37). Rotterdam, Netherlands: Sense Publishers.

Sinclair, A. (1988). La notation numérique chez l'enfant. In H. Sinclair (Ed.), *La production de notations chez le jeune enfant* (pp. 71–98). Paris, France: Presses Universitaires de France.

Sinclair, A. (2005). Las matemáticas y la imitación entre el año y los tres años de edad. *Infancia y Aprendizaje, 28,* 377–392.

Sinclair, A., & Sinclair, H. (1984). Preschool children's interpretation of written numbers. *Human Development, 3,* 174–184.

Sophian, C. (1996). *Children's numbers.* Boulder, CO: Westview Press.

Starkey, P., & Cooper, R. G. Jr. (1980). Perception of numbers by human infants. *Science, 210,* 1033–1035.

Starkey, P., Spelke, E. S., & Gelman, R. (1990). Numerical abstraction by human infants. *Cognition, 36,* 97–127.

Strauss, M. S., & Curtis, L. E. (1981). Infant perception of numerosity. *Child Development, 52,* 1146–1152.

Tolchinsky, L. (2003). *The cradle of culture and what children know about writing and numbers before being taught.* Mahwah, NJ: Lawrence Erlbaum Associates.

Van Loosbroek, E., & Smitsman, A. W. (1990). Visual perception of numerosity in infancy. *Developmental Psychology, 26,* 916–922.

Vygotsky, L. S. (1978). *Mind in society: The development of higher psychological processes.* Cambridge, MA: Harvard University Press.

Wynn, K. (1990). Children's understanding of counting. *Cognition, 36,* 155–193.

Wynn, K. (1992). Children's acquisition of the number words and the counting system. *Cognitive Psychology, 24,* 220–251.

Wynn, K. (1992). Addition and subtraction by human infants. *Nature, 358,* 749–750.

Wynn, K. (1996). Infants' individuation and enumeration of actions. *Psychological Science, 7,* 164–169.

CHAPTER 2

Young Children's
Self-Constructed Maps

Bárbara M. Brizuela and

Gabrielle A. Cayton-Hodges

In his research, Piaget (e.g., 1970, 1977) provided ample evidence for the ways in which children construct new understandings through an interactive process. In the interaction he described, new knowledge is not entirely contained in previous understandings, nor in the objects the learner interacts with, but in both, and in their interaction with one another. This chapter reports on an exploratory study in which we documented children's map productions while they produced and also interacted with different kinds of (conventional) maps and with their own self-constructed maps.

Central to our research is the premise that learning of symbolic systems such as maps is a slow, complex, and gradual process (Brizuela, 2004) in which the interaction with Piaget's *objects* (such as maps) is fundamental to the learning and development process. In fact, we believe that the absence of this interaction hinders learning and development and that it is fundamental to expose children to and provide them with opportunities to interact with and produce symbolic systems such as maps from an early age.

The Kindergarten children we worked with participated in a year-long series of activities (see Table 2.1) in which they interacted with conventional and self-constructed maps of different kinds. The sequence in Table 2.1 was designed to respond to different activities scheduled during the school year, such as field trips, as well as the regular classroom teacher's requests to address issues in both the children's mathematics and social studies curricula.

Our outline is in many ways similar to that presented by Liben, Kastens, and Stevenson (2002) and it also followed the methods Liben and Downs (2001) describe regarding the study of children's understanding of maps, focusing on production, comprehension, representational correspondence, and meta-representation (see Liben, 1999). Children were exposed to different types of maps, to different ways of representing the same space, and they themselves were also asked to produce different types of maps throughout the year. The types of maps and mapping tasks children were exposed to and were asked to produce fall into four main categories: (1) *diagrammatic maps* (e.g., draw the layout of your classroom); (2) *diagrammatic maps with guidance* (e.g., room blueprint provided and students detail the map); (3) *directional maps* (e.g., show how to get from the classroom to the library); and (4) *measurement and scale exercises* (e.g., which wall on the map will be the greatest number of inches).

Although the basic rationale for engaging in the exploratory study reported here is our belief in children's active construction of symbolic systems, there is further justification for this study in the research program of Liben and Downs. In their research, they argue for the "enthusiastic endorsement of early education in geography, in general, and of early education in mapping, in particular" (Liben & Downs, 1997, p. 164), stressing that education in mapping is not just exposure to maps but experiences with actual map-making.

We also took into account Lehrer and Pritchard's (2002) work with 8- and 9-year-old students learning about position and direction via mapping their playground. However, Lehrer and Pritchard's main focus was the "mathematizing" of the space. We conducted several lessons on measurement, but focused mostly on children's incorporation of landmarks, their production of maps that could be used by another individual for way finding or location understanding, and their understanding of the use of a map in general. In this chapter, we ask the question: What can we say about children's production of maps while they interact with these kinds of external representations?

CHILDREN'S UNDERSTANDING OF EXTERNAL REPRESENTATION SYSTEMS

Learning any external system of representation is a complex process. Martí (2003) describes these systems as realities organized around certain *formal properties* (hence, they are *systems*) that constitute *directly observable objects* (hence, they are *external*). Martí further highlights the choice of the terminology of *representation* versus *notation* or *symbol*, because the word implies both a product and a process. In this chapter, whenever we use the term "*representation,*" we mean external representation. This learning process is neither simple nor automatic; it is neither the result of mere experience nor of the activation of innate abilities or structures (Liben, 1981; Liben & Downs, 1991).

Research on children's learning of representations in mathematics and written language confirms that the interaction with conventional external representations and the production and subsequent interaction with self-constructed external

Table 2.1. Summary of Mapping Lessons

	Lesson	Description of Lesson
1.	Introduction	*Introduction:* Places we know and locating them on a globe.
2.	Pre-task	*Individual task:* Show the route you take from home to school.
3.	Map interpretation	*Exploring maps and map symbols:* After a field trip to the arboretum, we discussed what is important to put on a map of the place visited, and then looked at three different professional maps. How can we have three different maps of the same place? Discussion of the different elements included in a map. Drawing individual arboretum maps and discussing similarities/differences in one another's maps.
4.	Map interpretation	*Exploring maps:* Showed another map of the arboretum and discussed how we know it is a map. Why include a compass? What places in the arboretum were put on the map? What things are missing that we think should be on the map and why?
5.	Scale and measurement	*Map scale:* How does such a large place fit on a piece of paper? Introduction of the words *size, scale, left, right.* Everyone looks at the scale and compass on the map and brainstorms what they are for.
6.	Map production	*Planning a route:* Children take grid paper and make their own routes from their classroom to different locations in the school building (one group does the gym, another the auditorium, another the library, another the school office).
7.	Map interpretation	*Following a route:* Children follow their own maps of the school and discuss what they would change and what was easy/hard to follow.
8.	Map production	*Field trip to the park:* Children bring cameras and notebooks to take pictures of things that might be important in constructing a map.
9.	Map production	*Map construction:* Children each get their notebook and a piece of grid paper to draw their own maps to the park.
10.	Map interpretation	*Exploring maps:* Look at a map of their city and figure out the route we took to and from the park. Is it okay to rotate the map?
11.	Map interpretation	*Exploring maps:* Wrap up the semester by bringing in a globe. Where are we going for winter break and how long does it take to get there?
12.	Scale and measurement	*Map scale and measures:* Children talk about different ways to measure distances. One way is with our feet. Children are split into three groups and each group measures to a different place in the hallway. Is the distance the same in both directions?

Table 2.1. Summary of Mapping Lessons (continued)

13.	Scale and measurement	*Map scale and measures:* Last week's measurements are presented. Can we tell where each group measured to based on the numbers? Children are given the mode distance and draw diagrams of what the numbers correspond to.
14.	Map production	*Map perspective:* Children discuss perspectives and then draw a map of the classroom.
15.	Map production	*Plan construction:* Children draw another map of the classroom, this time with an outline of the walls to help.
16.	Scale and measurement	*Map scale and measures:* Children are put into groups and are given a tape measure. Measure each wall of the classroom, predict which walls you think will have the biggest/smallest numbers.
17.	Scale and measurement	*Map scale and measures:* Children are given an outline of the classroom and their measurements from last class. Which wall matches with which number? Paste the numbers on the map, then fill in the details.
18.	Map interpretation	*Exploring maps:* Look at a map of the city they live in. How many inches is it from the Amigos School to: the library? H. Square? C. Square? City Hall? What can we learn about the real distances from the map distances?
19.	Map interpretation	*Exploring maps:* Next week you are going to the museum. Look at the map of the city we live in and guess how long it will take to get to the museum. Use the places you know on the map to help you.
20.	Map interpretation	*Exploring maps:* Look at a map with all of the children's homes marked. Match the time it takes the child to get to school with the location on the map.
21.	Map production	*Map construction—Landmarks:* Make a list or draw pictures of all of the landmarks we see on the way to school.
22.	Map production	*Route construction:* Use the list you made last week to help you make a map of your route to school or your route to the bus stop.
23.	Map interpretation	*Exploring maps:* Next week we are going to the zoo. Look at a map of the zoo and show the route you will take to your favorite animal. How long do you think it will take to get there? What will you pass on the way?
24.	Map interpretation	*Exploring maps:* Compare the city you live in and the neighboring city. What do you see in each city? How are the grids of the cities different? What are the differences and similarities between the two maps? What does this tell us about the cities?
25.	Post-task	*Individual task:* Show the route you take from home to school.

representations *does* impact children's underlying understandings of the concepts embodied by the representations (e.g., Alvarado & Ferreiro, 2000; Brizuela, 2004; Ferreiro & Teberosky, 1979; Martí, 2003; Olson, 1994). In the field of mapping, Liben and Downs (1991) explain how, for instance, given the modified Mercator projection used in most two-dimensional maps, many people assume Greenland to be much larger in size than Saudi Arabia, when in fact they are approximately the same size. We would argue that this is an example of the enormous impact of interacting with external representations—as opposed to "direct experience in these places" (p. 148), as Liben and Downs (1991) point out, highlighting the interactive subject-object process that Piaget provided evidence for (e.g., 1970) in other areas of development. In our work, we purposefully position ourselves in a Piagetian framework because of his constructivist perspective and his rejection of nativist assumptions toward human development and learning. In this chapter, we address how young children gradually develop understandings of maps, as reflected in their map productions, and how their productions shift over the course of a school year.

CHILDREN'S DEVELOPMENT OF SPATIAL COGNITION

External representations that children produce are obviously impacted by their spatial cognition; these representations, at the same time, cannot be thought of as a complete reflection of spatial cognition. Piaget's constructivist theory provides the main framework for our chapter and it also provides grounding for our approach toward children's developing spatial cognition and understanding (see Liben, 1999). From Piaget's perspective (Piaget & Inhelder, 1948/1956; Piaget, Inhelder, & Szeminska, 1960), spatial understanding is constructed through interaction with the environment. Piaget and his colleagues reported on a number of studies dealing with children's perceptual, drawing, and mapping skills. Piaget and Inhelder (1948/1956) described three periods of spatial perception prior to representation. It is only in the last of these three periods, according to Piaget, that the child deals with the relationships of objects to one another, drawing begins, and space has begun to be not only purely perceptual, but also partly representational. Piaget (Piaget & Inhelder, 1948/1956) also reported on a series of stages in children's development of their diagrammatic layouts that finalize with the ability to construct an abstract plan with metric coordination, mastering distances and proportion.

There are some underlying differences that we hold with Piaget's perspective, however. Missing from Piaget's perspective, we believe, is the acknowledgment that spatial cognition is not constructed in isolation of interaction with external representations for space—his emphasis was on interaction with the environment. That is, while Piaget holds representation as being an end-state, perhaps alluding to a "conventional" or adult-like representation, we believe that representation is not only woven throughout each one of the stages but that additionally, representations interact in an intricate way with "spatial cognition" from the earliest stages.[1]

As Liben (1981) has pointed out, Piaget's theory stands in contrast with theories that adopt empiricist and innatist positions. Our position, similar to Liben's (1981) and Piaget's (Piaget & Inhelder, 1948/1956; Piaget, Inhelder, & Szeminska, 1960), is neither empiricist (e.g., Blaut, 1997a, 1997b) nor innatist (e.g., Landau, 1986), but constructivist. By adopting a Piagetian framework, we share the perspective that spatial cognition is a slow, complex construction, *not* merely the result of experience or of actualizing existing structures and abilities.

After early interest in children's construction of maps (e.g., Piaget, Inhelder, & Szeminska, 1960), researchers became doubtful regarding what could actually be learned though children's drawn maps (e.g., Kosslyn, Heldmeyer, & Locklear, 1977; Siegel, 1981). More recently, however, maps have emerged again in research on the development of spatial cognition (e.g., DeLoache, 1987; Landau, 1986; Liben & Downs, 1986; Presson, 1982, 1987). Uttal and Wellman (1989) have shown, for example, impressive map-reading abilities in preschoolers, highlighting that maps may be more useful to young children in their acquisition of spatial information than previously suspected. Uttal (2000) has encouraged us to adopt an interactive, constructive perspective by highlighting that "maps may alter how we think about the represented information" (p. 247), underscoring the perspective of Olson (1994).

OUR INVESTIGATION

The study described in this chapter is based on the following assumptions: Alvarado and Ferreiro (e.g., Alvarado & Ferreiro, 2000) have obtained crucial data from asking children "to write what they haven't yet been taught to write." In our study, what children "should or shouldn't know" did not limit our design; our goal was to explore the limits and possibilities of their understandings, and not what they may have possibly been taught. Thus, we assumed that asking children to construct their own maps, even though they "don't know" how to construct (conventional) maps, would prove very fruitful.

We describe children's self-constructed maps and identify what shifts in their self-constructed maps could be observed over the course of a school year. This study was exploratory in nature as we investigated the changes in children's representations over the school year. We do not, therefore, intend to make widespread claims about the generalizability of the results but merely show the extraordinary progress that these children made and highlight this as an example of enriching activities that show promise in curricula for young children.

The children with whom we worked were drawn from a two-way bilingual Spanish/English Kindergarten class in an urban area outside of Boston,

Massachusetts, that is racially, culturally, and ethnically diverse. Seven were male, 11 were female, and they ranged in age at the start of the school year from 4.6 to 6.5 years. The project was carried out during class time and was incorporated into both the social studies and mathematics curricula.

The Mapping Activities

Throughout the school year, all children participated in a weekly activity for approximately 1 hour in which we interacted with maps and different aspects of maps such as perspective, scale, and direction (see Table 2.1). A typical class consisted of approximately 30 minutes of group lesson/activity followed by 30 minutes of individual or small-group work, from which our data were derived.

There were three broad groups of activities in which we engaged during our year-long work with the children:

1. Measurement and Scale: The activities related to scale and measurement began with exploring how these are used in conventional maps. They also involved measuring distances ourselves, figuring out how to show these on paper, making relationships between distances measured and distances shown on a map, and estimating travel times by comparing distances on maps (relating time and distance). As part of this set of activities, we measured distances within the classroom (e.g., "How long are each one of our classroom walls?"), the school (e.g., "How far is it from our classroom to the bathroom?"), from home to school (e.g., "How far is my house from the school?"), and to different sites within the city that the children were visiting throughout the year (e.g., "How far is it from our school to the museum?").

The following were some of the highlights of the activities and examples of the kind of thinking that we could see during the activities:

A. *Related to measurement*: The children stated clearly and explicitly that one way to know how long something is was to "count with numbers." During Lesson 12, for instance, when Bárbara asked how to figure out the length of something, the following exchange occurred:

Callie[2]: With numbers.
Sara: We could put little lines the whole way down.
Zyllis: (Put lines) next to each other?

During Lesson 12, it was clear that the iss.ue of units of measurement and scale had not been fully developed by the children, with many children assuming that something is as long as however many tallies can fit (i.e., if ten tallies fit, then the stated length is "ten"). During this lesson, we discussed using "hands," "feet," and "inches" to measure distances and length. By the end of this lesson, children explicitly stated that "longer distances are associated with bigger numbers."

B. *Related to scale*: During Lesson 18, children began to explore issues of scale. For instance, when examining distances on a map of the city they live in, one child said, "In real life it's farther, they made the map smaller than in the real life." During Lesson 19, another child stated, "It is smaller [on the map] 'cause if you made it as it is it wouldn't fit." During this same lesson, a third child said, "Sometimes when we look at something it looks close to each other but it's really far away."

2. Map Interpretation: Even though our focus in this chapter is on map production, during our work with the children we carried out many map interpretation activities. Our goal with this kind of activity was to provide children with ample opportunities to interact with different kinds of maps (e.g., of their city—in different scales and formats; of the museums they would be visiting; of the zoo; of the world). We also sought to spend time on the analysis of specific elements (landmarks, rivers, schools, monuments, public buildings) that are represented in maps. Our activities included interpretation as well as production elements, always trying to combine both. Further, we always attempted to make connections between children's lived experiences of routes taken and how these translate into representations on maps.

The activities included working with a map of the city they live in, looking for what children could see and recognize, recalling important landmarks and trying to find them on the map, measuring distances on maps, and comparing distances on maps. In working with maps of the city in which each one of their homes was marked, for instance, the following exchange occurred:

Bárbara: Who do you think probably takes the longest to get to school?
Norman: Callie 'cause she lives farthest away [from the school].
Bárbara: Who probably takes the shortest amount of time?
Norman: Myself! [He lives one block from school.]

The focus of activities was on identifying potential routes, and identifying landmarks they knew or that seemed important in maps of the city they live in or sites they were going to or had recently visited.

3. Map Production: Besides interacting with different kinds of maps, children were also asked to produce different kinds of maps, the main focus of this chapter. Once again, the focus was on the production of routes and landmarks related to their homes, their city, their school, and places or sites they would be visiting (the park, for instance, as will be seen below). During Lesson 21, for instance, children made lists or drawings of the landmarks they saw on the way to school every day. Subsequently, during Lesson 22, children were asked to draw the route from home to school, using the work they had done during Lesson 21 as an aid, as well as maps of the city that hung throughout their classroom.

Analysis of Children's Work

Our analysis for this chapter centers on children's work on four production tasks:

1. Two implementations (September 2004 and June 2005) of the task "Show the route you take from home to school" presented in the context of an individual interview.

2. Two implementations (November 2004 and June 2005) of the task "Show the route you take from your school to the neighborhood park," which were carried out following a walk to the park, as an individual production activity carried out in the classroom.

Our analysis of children's work on these tasks centers on the qualitative differences observed among the children's work, most notably from the beginning to the end of the school year. The analysis does not count correct or incorrect responses. Rather, at the beginning of the school year, we observed that most of children's productions were drawing-like, and seemed more like lists of the objects seen on their routes rather than actual directional maps. Wolf and her colleagues (e.g., Perry & Wolf, 1986; Wolf & Gardner, 1985) have shown that between 5 and 8 years of age, children are able to distinguish between maps and drawings. Therefore, one of the themes we considered in our analysis was:

Could children's productions be differentiated into drawings and maps? Children's productions were considered to be more like a map if they included lines, arrows, and simple drawings as part of a route construction. They were also considered more like maps if they were more schematic, as opposed to full figurative drawings. All classification of students' written productions was carried out by consensus between the chapter authors.

Because our research questions centered on the shifts and changes in children's map productions, our second set of categories dealt with what kinds of conventions we observed children gradually incorporating into their map construction. Furthermore, Lehrer and Pritchard (2002) highlight that an important means of testing conjectures about "good" maps would lie in students' knowledge of the landmarks and general configuration of this familiar space. Therefore, a second theme we considered in our analysis was:

Did children incorporate conventions in their maps? Specifically, the map: (a) shows ending and/or starting points symbolically, as a cross or dot; (b) shows route; and (c) shows landmarks along a route.

FINDINGS

In general, more children constructed map-like productions at the end of the school year (from 58.82% to 77.78% in the home to school route task, and from 0% to 58.82% in the school to park route task), and more mapping conventions—specifically, showing ending and/or starting points as cross or dot, showing a route, and showing landmarks along a route—were incorporated into their productions as well. The following are examples of the themes observed among children's productions in each of the four production tasks.

Differentiation Between Drawings and Maps

Comparison of Responses to: "Show the Route You Take from Home to School"

Zyllis (female; 5y,4m³ in September)

In Zyllis's two productions from September to June, we have an example of a drawing, on the one hand, of herself and her dad in Figure 2.1, and of a route between two locations on the other hand (Figure 2.2). In Figure 2.2, Zyllis names and draws different landmarks along the route. Her end-of-year production includes lines and simple drawings incorporated into a route, as opposed to a full figurative drawing.

Figure 2.1. Zyllis Shows the Route from Her House to School in September 2004. On the Left is "the Route," and on the Right the Van with Her Dad and Herself in it.

Figure 2.2. Zyllis Shows the Route from Her House to School in June 2005. On the Left is Her House; the Top Horizontal Line Indicates a Tunnel, the Lower Horizontal line "a Cave," and on the Bottom Right is the School.

Walk to Park in the Fall and the Spring

Yamila (female; 6y,7m in November)

Yamila's production at the end of the school year (see Figure 2.4) is, like Zyllis's, more schematic than it was at the beginning of the school year (see Figure 2.3). While at the beginning of the school year her production centered on an inventory of the objects and landmarks observed along the route to the park, at the end of the school year these objects and landmarks are integrated into a path from the school, the point of origin of her route, to the destination, the park (see Figure 2.4). We considered this integration to indicate her appropriation of conventional mapping "rules."

Shows Ending and/or Starting Points as Cross or Dot

Comparison of Responses to: "Show the Route You Take from Home to School"

Abril (female; 5y,10m in September)

Although both of Abril's productions are relatively similar (her later production, shown in Figure 2.6, includes, however, an additional leg to the trip), in her later map she includes a star to indicate the end point of her route, perhaps beginning to consider some of the conventions used in maps such as those she had interacted with in class. In her later production she does not include the name

Figure 2.3. Yamila Shows the Route to the Park in November 2004

Figure 2.4. Yamila shows the route to the park in June 2005. She describes the lower vertical line saying "go straight," the horizontal line saying "turn left," and the top vertical line saying "then go straight and there is the park."

of the street that she lives on, which she had done at the beginning of the school year (see Figure 2.5). In this way, she may be emphasizing information other than street names—such as relative locations and landmarks—that is relevant in map productions.

Walk to Park in the Fall and the Spring

Hannah (female; 5y,5m in November)

At the end of the school year, Hannah indicates the destination of her route—the park—with an "x." Additionally, she shows the objects she sees no longer as isolated (see Figure 2.7), but as landmarks *along* a route (see Figure 2.8). This latter change is similar to that observed in Yamila's productions in Figures 2.3 and 2.4.

Figure 2.5. Abril Shows the Route from Her House to School in September 2004. On the Left is Her House, and on the Top Right is a Street Light at the Corner of the School, and the School.

Figure 2.6. Abril Shows the Route from Her House to School in June 2005. At the Top of the Right Vertical Line is "My House," the Bottom Horizontal Line is Showing "Grass," and the Star is Indicating the School.

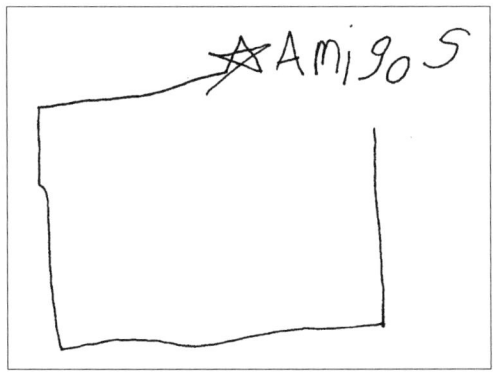

Shows Route

Comparison of Responses to: "Show the Route You Take from Home to School"

Sara (female; 5y,3m in September)

As other children did, Sara began by producing a drawing of herself in her mother's car at the beginning of the school year (see Figure 2.9). By the end of the school year, she was able to produce a route, showing the three stops along the route (see Figure 2.10).

Walk to Park in the Fall and the Spring

Callie (female; 5y,7m in November)

Just as the examples described, Callie begins by making isolated drawings of objects and landmarks she sees along the path to the park (see Figure 2.11). By the end of the school year, she constructs a route, instead of a drawing (see Figure 2.12). In this route, she only includes an "x" to mark the route destination, and the names of the streets taken, but does not integrate the kinds of objects and landmarks drawn in Figure 2.11.

Figure 2.7. Hannah Shows the Route to the Park in November 2004. At the Top Left is an Airplane, Below that a Building, a Bridge to the Right of the Building, and on the Right-hand Side a Tree and the Sun.

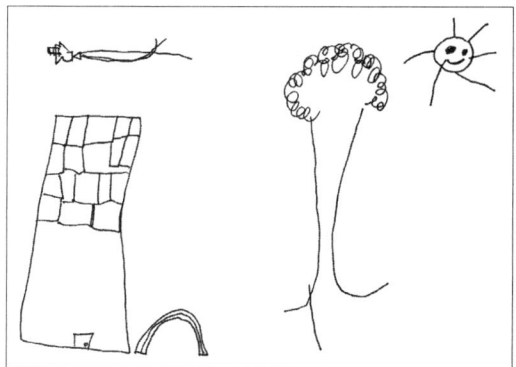

Figure 2.8. Hannah Shows the Route to the Park in June 2005. From Top to Bottom She Shows Her School, a Stop Sign, Her House, a Car, Another House, and at the Bottom "We Arrived at the Park."

Figure 2.9. Sara Shows the Route from Her House to School in September 2004. She Shows Herself and Her Mom in the Car, as well as the Car's Steering Wheel on the Left.

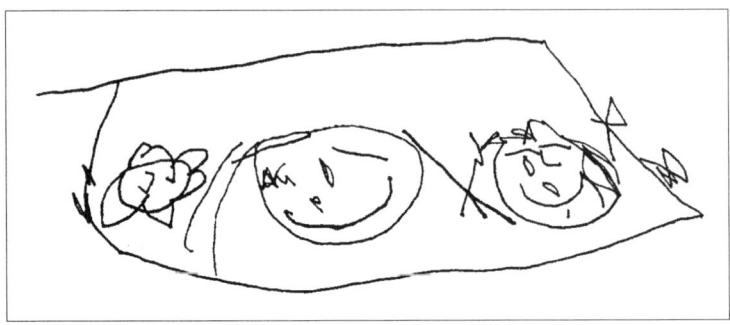

Figure 2.10. Sara Shows the Route from Her House to School in June 2005. From Left to Right She Indicates Her Home, Her After-school Program, and the School.

Figure 2.11. Callie Shows the Route to the Park in November 2004

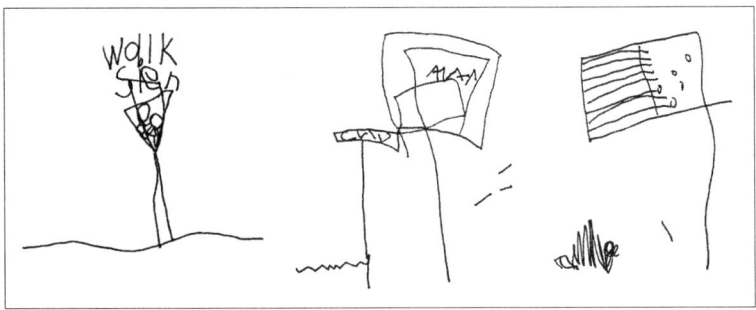

Shows Landmarks Along a Route

Comparison of Responses to: "Show the Route You Take from Home to School"

Callie (female; 5y,7m in November)

From the beginning (see Figure 2.13) to the end of the school year (see Figure 2.14), Callie has begun to integrate objects she sees along the way on the route taken from home to school. In Figures 2.11 and 2.12 we see her production for a different route, in which she does not integrate the objects and landmarks seen into the route. It is interesting to note that Callie does show a much more schematic route when depicting the route from the school to the park. This is likely due both to the far distance from her house to the school (which she traveled on

Figure 2.12. Callie Shows the Way to the Park in June 2005. She Shows the Names of Streets She Sees on the Way to the Park.

Figure 2.13. Callie Shows the Route from Her House to School in September 2004. On the Left-hand Side is the School, and on the Right is Her House.

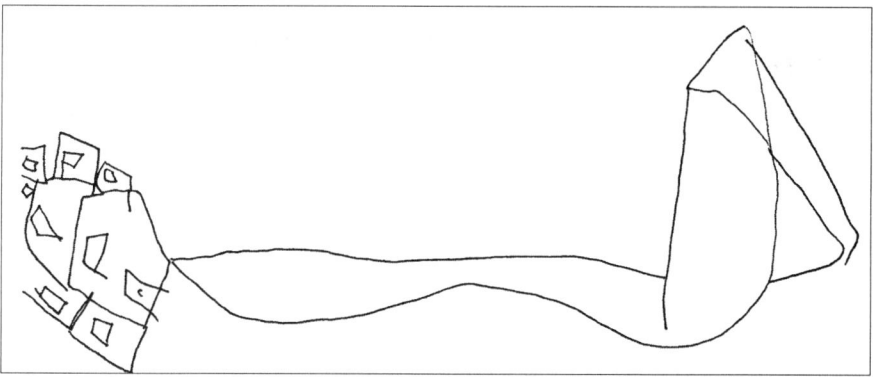

a school bus; see the comment above from the children who indicate that Callie lives farthest away from the school), as opposed to the short walking route that the class took to the park, which may have allowed her more opportunities to actually see more landmarks and objects, as well as her greater familiarity with the landmarks on the home–school route due to traveling it twice every day.

Walk to Park in the Fall and the Spring

Justin (male; 6y,5m in November)

Once again, as seen in other examples above, Justin begins by making a drawing at the beginning of the school year (see Figure 2.15); by the end of the school year (see Figure 2.16), he is able to integrate simple drawings into the route taken to the park, along with directions and indications of other landmarks.

DISCUSSION

The results presented indicate that, while children's productions of external representations of space are far from expert-like, accurate, and conventional, they have changed from the beginning to the end of the school year. By the end of the school year, their maps still lack scale, perspective, and proportion, for instance. Still, they are both qualitatively and quantitatively different from what they were at the beginning of the school year. Furthermore, the children were not specifically trained to produce maps for these *particular* routes. These were spaces that the children had previous experience with (although with different degrees of familiarity, as we saw in Callie's example above), and in this sense were natural, known environments for them (as opposed to an artificial laboratory, for instance).

From the beginning to the end of the school year, more children made map-like representations for routes than at the beginning of the school year. We also observed in Callie's case that familiarity and access to more objects that she could directly observe led to her production of different kinds of maps—less detailed

Figure 2.14. Callie Shows the Route from Her House to School in June, 2005. She Shows Her Home on the Left and the School on the Right, as well as Other Buildings on the Path Between Both.

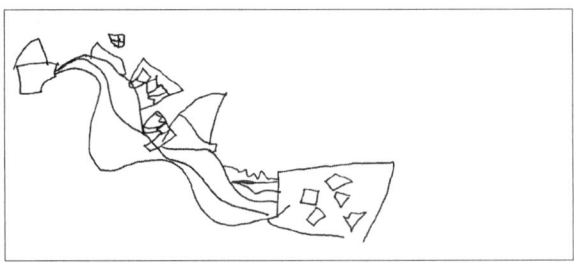

Figure 2.15. Justin Shows the Route to the Park in November 2004. He Shows an Airplane at the Top, and a Cloud and the Sun Below it. On the Left He Shows a Building, a Tree on the Right, and Grass at the Bottom.

Figure 2.16. Justin Shows the Route to the Park in June 2005. Starting at the School at the Bottom Right, He Shows, Going Left, a Fence and a Tree, then Indicates on the Vertical Line "Go Straight," then "Turn," then "Go Straight," then "Turn Again," then "Go Straight Again," and Finally, at the Circle, the Park.

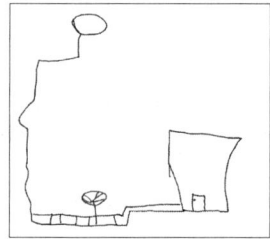

and more schematic in the case of a considerably shorter route with which she had less familiarity. In addition, at the end of the school year, more children showed starting and/or ending points as dots or crosses, showed routes, and showed landmarks *along* a route, beginning to incorporate some of the conventions they had interacted with in the conventional maps they worked with during the school year. We would also like to note that the two tasks presented (route from school to park and route from home to school) were similar tasks for children who walked to school, yet quite different tasks for children who lived far from the school and traveled via bus or car. In some cases, children traveled for 30 minutes to school and thus the route was rather complex, with a more passive role in the travel.

Some could argue that these observed differences are merely the effect of age, growth, or even possibly of a better comprehension of the instructions given to them in the different tasks (i.e., "Show the route you take from a to b"). Although we are unable to definitively speak toward any of these arguments, given the exploratory nature of the study, we would like to argue that exposure to maps, mapping, and interaction with them did have some impact on the children. That is, the work with the children was not fruitless. Additionally, given the large age range within the class, we were able to compare the younger children at the end of the year with the older children at the start of the year as an age comparison. It was clear that the progression we saw was not simply due to age, as the older children were not producing more sophisticated maps than the younger children.

The observed differences in children's map productions provide evidence for the impact of interacting with external systems of representation—in this case, maps—as well as with the mapping process. Interacting with maps is an important part of learning how to make maps, understanding them, appropriating their conventions, and developing an understanding of the rules that underlie their functioning. This constructive process cannot occur in the absence of the interaction with the very objects children are being asked to learn about. We also want to underscore the role that may have been played by children's reflective abstraction (Piaget, 2001) on maps and mapping—what Liben (1999) calls meta-representation—through the many opportunities children were provided to reflect on the meaning and role of different aspects of maps and mapping.

Our results indicate the value and importance of incorporating maps and mapping into early childhood education. Our results illustrate that interaction with representational objects has an impact on children's ability to represent. While the results observed are surely not merely the effect of this interaction, they do highlight the need to know more about children's spontaneous abilities to construct maps as part of their development. The results also indicate, in general, the need for more work to uncover aspects of students' grappling with the complexity of learning a conventional system.

Still, many questions remain, which we look forward to addressing in future studies: What role does familiarity with a particular route or space have on children's productions? How can this familiarity be used to facilitate their interactions with the conventions of mapping? How accurate do children's productions become over time? What kinds of correspondences (representational and geometric) are gradually incorporated into children's productions (such as accurate routes, schematic routes, and so on)? How are they incorporated and appropriated?

NOTES

1. Some might even argue (Nemirovsky, 2009) that separating "cognition" from "representation" is arbitrary and fruitless. From this perspective, which we share, though perhaps not so radically, we might begin to wonder what is a representation without spatial cognition as well as what is spatial cognition without representation.

2. All names used in this chapter are pseudonyms.

3. In this chapter, age is denoted as (years,months).

REFERENCES

Alvarado, M., & Ferreiro, E. (2000). El análisis de nombres de números de dos dígitos en niños de 4 y 5 años [The analysis of names of two-digit numbers in 4- and 5-year-old children]. *Lectura y Vida. Revista Latinoamericana de Lectura, 21*(1), 6–17.

Blaut, J. M. (1997a). The mapping abilities of young children: Children can. *Annals of the Association of American Geographers, 87*(1), 152–158.

Blaut, J. M. (1997b). Piagetian pessimism and the mapping abilities of young children: A rejoinder to Liben and Downs. *Annals of the Association of American Geographers, 87*(1), 168–177.

Brizuela, B. M. (2004). *Mathematical development in young children: Exploring notations.* New York: Teachers College Press.

DeLoache, J. S. (1987). Rapid change in the symbolic functioning of very young children. *Science, 238*, 1556–1557.

Ferreiro, E., & Teberosky, A. (1979). *Los sistemas de escritura en el desarrollo del niño [Literacy before schooling].* México City, México: Siglo XXI.

Kosslyn, S. M., Heldmeyer, K. H., & Locklear, E. P. (1977). Children's drawings as data about internal representations. *Journal of Experimental Child Psychology, 23*, 191–211.

Landau, B. (1986). Early map use as an unlearned activity. *Cognition, 22*, 201–223.

Lehrer, R., & Pritchard, C. (2002). Symbolizing space into being. In K. Gravemeijer, R. Lehrer, B. van Oers, & L. Verschaffel (Eds.), *Symbolization, modeling and tool use in mathematics education* (pp. 59–86). Dordrecht, Netherlands: Kluwer Academic Press.

Liben, L. S. (1981). Spatial representation and behavior: Multiple perspectives. In L. S. Liben, A. H. Patterson, & N. Newcombe. (1981). *Spatial representation and behavior across the life span. Theory and application* (pp. 2–36). New York: Academic Press.

Liben, L. S. (1999). Developing an understanding of external spatial representations. In I. E. Siegel (Ed.), *Development of mental representation: Theories and applications* (pp. 297–321). Mahwah, NJ: Lawrence Erlbaum Associates.

Liben, L. S., & Downs, R. M. (1986). *Children's production and comprehension of maps: Increasing graphic literacy* (Report to the National Institute of Education; Grant No. 6-83-0025). Washington, DC: National Institute of Education.

Liben, L. S., & Downs, R. M. (1991). The role of graphic representations in understanding the world. In R. M. Downs, L. S. Liben, & D. S. Palermo (Eds.), *Visions of aesthetics, the environment, and development: The legacy of Joachim F. Wohlwill* (pp. 139–180). Hillsdale, NJ: Lawrence Erlbaum Associates.

Liben, L. S., & Downs, R. M. (1997). Can-ism and can'tianism: A straw child. *Annals of the Association of American Geographers, 87*(1), 159–167.

Liben, L. S., & Downs, R. M. (2001). Geography for young children: Maps as tools for learning environments. In S. L. Golbeck (Ed.), *Psychological perspectives on early childhood education: Reframing dilemmas in research and practice* (pp. 220–252). Mahwah, NJ: Lawrence Erlbaum Associates.

Liben, L. S., Kastens, K. A., & Stevenson, L. M. (2002). Real-world knowledge through real-world maps: A developmental guide for navigating the educational terrain. *Developmental Review, 22,* 267–322.

Martí, E. (2003). *Representar el mundo externamente: La adquisición infantil de los sistemas externos de representación* [*Representing the world externally: Children's acquisition of external systems of representation*]. Madrid, Spain: A. Machado Libros.

Nemirovsky, R. (2009). Remarks on external representations as learning tools. In C. Andersen, N. Scheuer, M. P. Pérez Echeverría, & E. Teubal (Eds.), *Representational systems and practices as learning tools* (pp. 281–296). Rotterdam, Netherlands: Sense Publishing.

Olson, D. (1994). *The world on paper: The conceptual and cognitive implications of writing and reading.* Cambridge, MA: Cambridge University Press.

Perry, M. D., & Wolf, D. (1986, May). *Mapping symbolic development.* Paper presented at the 16th Annual Symposium of the Jean Piaget Society, Philadelphia, PA.

Piaget, J. (1970). Piaget's theory. In P. H. Mussen (Ed.), *Carmichael's manual of child psychology* (3rd ed.), pp. 703–732. New York: John Wiley & Sons.

Piaget, J. (1977). *La naissance de l'intelligence chez l'enfant.* Paris, France: Delachaux et Niestlé.

Piaget, J. (2001). *Studies in reflecting abstraction* (R. L. Campbell, Ed. & Trans.). Philadelphia, PA: Psychology Press. (Original work published 1977)

Piaget, J., & Inhelder, B. (1956). [1948]. *The child's conception of space* (F. J. Langdon & J. L. Lunzer, Trans.). New York: W. W. Norton & Company.

Piaget, J., Inhelder, B., & Szeminska, A. (1960). *The child's conception of geometry.* New York: W. W. Norton & Company.

Presson, C. C. (1982). The development of map-reading skills. *Child Development, 53,* 196–199.

Presson, C. C. (1987). The development of spatial cognition: Secondary uses of spatial information. In N. Eisenberg (Ed.), *Contemporary topics in developmental psychology* (pp. 77–112). New York: Wiley.

Siegel, A. W. (1981). The externalization of cognitive maps by children and adults: In search of ways to ask better questions. In L. S. Liben, A. H. Patterson, & N. Newcombe (Eds.), *Spatial representation and behavior across the life span: Theory and application* (pp. 167–194). New York: Academic Press.

Uttal, D. H. (2000). Seeing the big picture: Map use and the development of spatial cognition. *Developmental Science, 3*(3), 247–286.

Uttal, D. H., & Wellman H. M. (1989). Young children's representation of spatial information acquired from maps. *Developmental Psychology, 25*(1), 128–138.

Wolf, D., & Gardner, H. (1985). *Broadening literacy: A final report for the Carnegie Corporation.* Cambridge, MA: Harvard Graduate School of Education.

CHAPTER 3

Kindergartners' Idiosyncratic Representations of Movement

Jason Kahn

Idiosyncratic, external representations of movement are a valuable lens into the thinking of young children. Movement is the coordination of two variables at the same time: distance and time. This observation is relevant to the development of understanding of motion (Piaget, 1970) and in explaining findings that building sophisticated understanding of movement is challenging for all learners (e.g., McDermott, Rosenquist, & van Zee, 1987). Young children's idiosyncratic representations give us a richer understanding of the development of an important scientific concept. They also let us look deeper into how children use representations to frame and extend their understandings. Because movement is so present across their lives, their representations of motion are rich and dynamic canvases that give hints at the mechanisms the children use to build upon their own knowledge. This chapter asks how children enrolled in Kindergarten produce representations of movement, and it teases apart the mechanisms that they use to convey meaning in their representations.

When children are asked to show their thinking as a representation, they are externalizing their thoughts. An external representation (Martí & Pozo, 2000) is an expression that one simply creates outside the body, such as a drawing. An external representation does not have to be shared with others, as it can simply be an aid to help structure thought. In considering the idea of looking at children's external representations, it is important to remember that the line between internal and external is likely a false one. Nemirovsky and Noble (1997) and Nemirovsky and Ferrara (2009) argue that internal and external representations feed each other to the point that any claim of dualism is meaningless. Similarly, Ochs, Gonzalez, and Jacoby (1996) found in ethnographic research that scientists came to embody their representations, making very little distinction in their professional communication between where the representation ended and the individual thought began. When we acknowledge that there is no bright line between an internal and external representation, we can see produced external representations, be they drawings, gestures, or utterances, as a small clue into the thinking of the person who produced it. This study sets out to collect these clues and illustrate how young children take opportunities to create external representations to show their thinking about movement.

Children's drawings are idiosyncratic, meaning that they do not superficially follow a conventional system of rules that commonly governs traditional notational systems (Goodman, 1976). A notational system—such as writing—is a highly and specifically defined way of communicating a set of information. Motion graphing can be considered a notational system when the set of functions is deliberately limited to highly discrete cases, such as moving away versus moving closer, or moving fast versus moving slowly. When we think about the discrete nature of a notational system, it is important to consider that children can see their own idiosyncratic representations as highly defined. Karmiloff-Smith (1990) noted that Kindergarten-aged children strenuously resist any attempt to mislabel their drawings. In the child's mind, a drawing is unambiguous, which is at least part of a well-defined notational system.

In the study of representations of movement that is described in this chapter, I gave the children a healthy amount of freedom to create representations that showed their own understanding of a particular concept. As it is reasonable to expect that children will produce drawings as representations of movement, it is also reasonable to expect the children to invent ways of representing motion that may seem surprising to adults. Invented representations (Brizuela, 1997; Danish & Enyedy, 2007; Sherin, 2000) refer to the creative process that children engage in when they are asked to create representations of a phenomenon they have either not previously considered explicitly (which is not likely in the case of movement), or they have not previously been asked to carefully consider how to represent (which is more likely in the case of movement).

Children will readily represent a number of phenomena, and previous work does directly discuss invented representations surrounding movement. Two studies are of particular relevance to this research. In one case, diSessa, Hammer, Sherin, and Kolpakowski (1991) followed middle school children through a sequence of activities related to movement, where they were expected to invent and critique a series of static representations of movement. The children had pre-existing resources to create representations, and they used a broad tool set that included, among other things, drawings, arrows, and writing. The children also had resources to consider and critique nonconventional representations they were being shown, such as an array of slanted lines to show speed, and in the context of an immersion activity were able to demonstrate their skills (diSessa et al., 1991; diSessa & Sherin, 2000). Meta-representational competence is more than an ability to interpret a conventional representation; it refers to an ability to interpret and manipulate a representation to help structure thought around a given concept, such as movement (diSessa

& Sherin, 2000). Strong meta-representational competence means that a learner is comfortable choosing the appropriate representation for a given problem and building novel representations should the situation call for it.

In a second case, Sherin (2000) looked at 6th-graders specifically tackling movement with a variety of invented, novel, and conventional representations of movement. In general, Sherin's data support the claim that students are well served by being able to manipulate a range of representations, and being given freedom to construct individual invented representations. He also specifically discusses some of the drawing strategies of the students, finding that the children are comfortable using drawing as a representational medium. This work also illustrates how children can utilize a drawing as a representation to show movement; the children in this case used space in the representation to correspond directly to space in the physical world. We could predict that younger children, such as those in this study, could also use their representations as a direct mapping of a motion phenomenon unfolding in front of them.

This brief context sets the scene for research related to young children's invented representations and their appropriation of conventional representations of movement. This research addresses the question of how Kindergarten-aged children present movement when asked to invent representations. In answering this question, the research breaks apart the elements of their representations and looks at how the elements work together to reveal the child's thinking about movement. Children are already building resources to understand movement (Hammer, 2000), and it seems likely that they are building resources to construct and manipulate representations (Sherin, 2000). However, little is known about young children's representations of movement, and this chapter explores this question by exploring children's representations of a specific motion task.

ELICITING CHILDREN'S REPRESENTATIONS OF MOTION

To elicit representations of movement, I interviewed ten kindergartners from laboratory schools at a private university in the Northeast. Both schools focus heavily on child-centered teaching.

The data discussed here focus on the kindergartners' productions. The productions centered on movement cues related to the motion of a toy car that either moved away, toward, or stayed still in relation to the child. The car moved fast, slowly, or not at all. Children had access to a car and a short track so they could present the motion in physical form before producing their representation on paper.

In the first part of the interview, both the child and the interviewer shared knowledge about what the car would do. The interviewer asked the child to consider the following four movements: (1) the car moving away slowly; (2) the car moving away quickly; (3) the car standing still; and (4) the car moving away,

changing directions, and returning. In each case, the child was given an opportunity to show the movement with the car on the track. After that, the child was
asked to "put something on paper" to show the car's movement. The child was
given as much time as needed. Wording the task in this manner avoids biasing the
form of the child's representation (as in not leading them to a drawing or writing).
After the child created the representation, the interviewer asked a series of questions structured to elicit information about the representation. The interviewer
asked questions about the elements of the representation and what they meant.
Being a clinical interview (Duckworth, 1973; Ginsburg, 1997), the main goal of
the interviewer was to facilitate the child's self-expression.

The second part of the interview followed a similar structure with one exception. A computer program would select a series of movements from a pool of possibilities. The pool included permutations of the car moving away from the child,
moving closer, moving quickly, moving slowly, and not moving. The child was
verbally presented the movement by a computer program (which read the motions
aloud to the child because some children could not read, and randomized what
movement the child heard), but the interviewer was not aware of the motion that
had been presented to the child. Because the interviewer did not know the movement, the child had to use the representation to communicate an unshared idea.
The child created three representations. In the first case, the car had no change in
its movement. In the second case, the car changed its direction or speed once, and
in the third the car changed its direction or speed twice. The movement presented
to the child was stored by the program to keep track of how motion and representations were matched. After the children created each representation, they were
interviewed as before.

SUMMARY OF THE CHILDREN'S REPRESENTATIONS

The children in the study were asked to show a car moving in relation to a person;
therefore, it is not surprising that many of the representations included the physical objects of a person and a car. While the children relied on the same building
blocks in their drawings, they adopted diverse strategies for showing the actual
movement of the car. This section describes both key elements of the children's
drawings: the items they chose to put on paper to establish a context for motion,
and the mechanisms they used to build a representation of the actual motion.

The Major Elements of the Representations: Cars and People

Given that the children were asked to create representations of a moving
car, it is not surprising that in 76 representations of movement, all but one child
drew an object that was either described as a car by the child, or clearly contained

elements to identify the drawing as a car, such as wheels, lights, or a steering wheel. However, while all the children drew cars, there was variation in how the cars were presented on paper. For example, Amelia produced the representation shown in Figure 3.1.

On the left of the representation is a person. We can understand this since the person has certain features that clearly identify it as a person: a head, eyes, a smile, hands, and feet. A common developmental assessment of children (Gardner, 1982) is to count the number of features in their drawings of people, and children typically add features to the people in their drawings throughout early childhood. On the right, there is a box. This box is a car, but the only way that this is confirmed is through the interview. The following exchange immediately precedes Amelia making the drawing in Figure 3.1.

Jason: So we drew the person . . . what would we draw next?
Amelia: A car!
Jason: Okay, do you want to draw a car?
Amelia: (Draws a box; see right-hand side of Figure 3.1)

Simple exchanges like the one above are invaluable in classifying children's drawings when the child forgoes identifiers in the drawings. On the other end of the spectrum are cars with considerably more detail. For example, in Figure 3.2, Nora draws what is clearly a car. The car has four wheels, a front delineated by headlights, a cockpit, a steering wheel, and a driver. The car is also surrounded by lines; as we explore these drawings on a deeper level than just the car itself, we will see that these lines represent and qualify the *movement* of the car.

Cars with a high level of detail, as in Figure 3.2, were not unusual. Forty-eight of the kindergartners' productions (64%) included an element that could clearly be interpreted as a part of a car. Conversely, in 27 of the 76 (36%) productions children drew a representation of a car with no identifiable car parts (like wheels, windows, and so forth). When comparing level of detail across children's different representations, the level of detail can be thought of as bringing a figurative element into the drawing (de Saussure, 1931; Piaget, 1929/2007), or a literal translation of the physical world into the drawn representation. Gardner (1982) saw children's increasing literalism in their drawings as an attempt to show mastery over the environment and the representation. Children's inclusion of detail in their drawing of cars shows that this may be the case. Children showed detail either to match the literalism of the prop cart, to present a faithful re-creation of cars that they would commonly see, or to demonstrate fanciful cars.

Thinking about how detailed children are when they present the basic elements of their drawings, people and cars, might seem to set the stage for how explicit the presentation of movement phenomena is. However, there are rare exceptions that should give us pause. In Figure 3.3, Rex has represented a car

Figure 3.1. Amelia's First Representation of a Moving Car (the Car is on the Right)

Figure 3.2. Nora Drew a Car With a High Level of Detail

moving away slowly. The representation has four parts. A horizontal line that divides the representation shows the track that the car moves along. A dot at the far right shows the end of the track. A dot above the line shows the car, and a dot below the line shows Rex, the child making the drawing. Rex explained his signifiers (de Saussure, 1931; the form used to express an underlying idea, or what is signified) in an interview, and acknowledged that others may not understand his drawing unless he explained it to them. However, he felt that his goal was to show the movement, and that he did not want to include too many details. In some ways, Rex's purposeful eschewing of detail seems to add a layer of sophistication to the representation rather than detract. He makes certain features implicit in order to focus his representation on communicating what he decided were the important elements of the representation.

Figure 3.3. Rex's Representation of a Car (the Dot Above the Line), a Track (the line), and a Person (the Dot Below the Line)

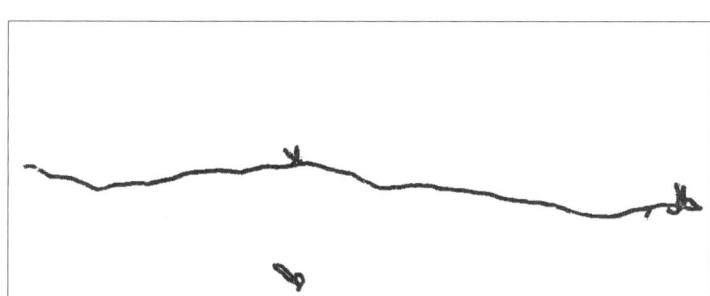

Representations make certain aspects of the phenomenon being represented explicit, while leaving other aspects of the phenomenon implicit (Pozo, 1999). For instance, some children decided to make the details of the car explicit. However, Rex decided that the car should be abstracted, and that the details of the car were not as important as developing a system for showing movement. Rex demonstrates how certain children will consciously choose to omit information for the sake of emphasizing other elements in their representation. He is capable of developing a representational economy, meaning that he weighs different elements for inclusion, and then selects those that are personally important. The task asked him to do more than simply represent a car (or a person), and he focused on a way to show movement.

When turning attention to representations of people, we start to get a sense of the system that children used to represent movement in their representations. A slight majority of the children's representations (39 of 76, or 51%) included a representation of a person. In the context of the task, the car's movement is always presented relative to the child. Therefore, the representation of a person could help establish a point of reference (or origin) for the representation. As the task was presented as "Please put something on paper to show a car moving [toward/away] from you," including a person is one possible way to establish the direction the car is moving. Children who did include people in their drawings either put the person physically touching or overlapping with the car (potentially as a driver), a person detached from the car (as a potential point of reference), or a combination of both person–car arrangements.

Nine representations (12%) included a person physically touching the representation of a car. Within these nine representations, all of the children identified the person as themselves and also as the driver of the car. Figure 3.4 shows an example of a person attached to the drawing of the car. Typically, when a person was attached to the car, the child did not see the person as adding to the depiction of movement phenomena, as the only exception is shown in Figure 3.5. The driver of the car is saying, "Wow! Wow!" to show that the car is moving fast.

Figure 3.4. James Drew Person Attached to the Car, as a Driver

Figure 3.5. The Driver Qualifies the Speed of the Car by Saying "Wow! Wow!" to Show the Car is Moving Fast. The "Wow! Wow!" is Found in the Bubbles Extending from the Driver in the Car.

Thirty representations (40%) included a drawing of a person spatially separated from the car. Children's drawing of people outside of cars was often to demonstrate a specific idea related to the movement of the car in the drawings. In the quote below, Duncan described his use of a person in Figure 3.6.

> *Jason*: Is it important to put (a drawing of) you in the picture? (See Figure 3.6.)
> *Duncan*: It's like . . . it's like so you could know what way the car is moving.

**Figure 3.6. An Example of a Drawing with Three People; "Duncan the Kid,"
the Interviewee; Jason, the Interviewer; and "Duncan the Guy," the Car
Driver.**

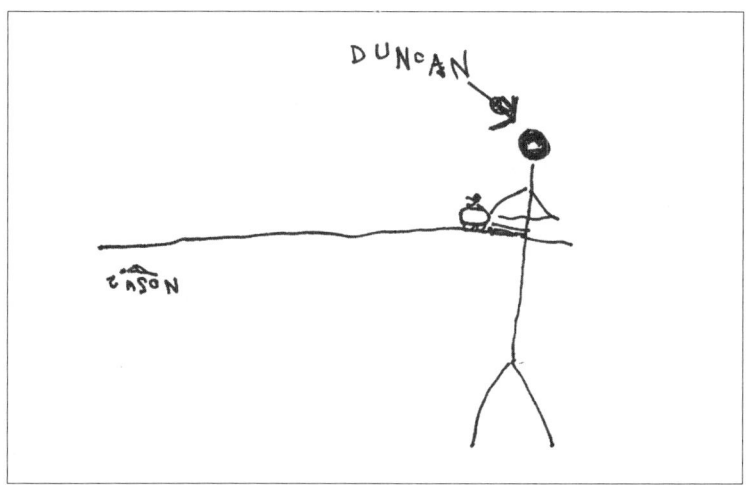

There are three people shown in Figure 3.6. Two are labeled, and Duncan is
talking about the figure that he labeled with his name. He has also included me
(labeled as Jason) and a rider (which he called "Duncan the guy," and is distinct
from what he called "Duncan the kid," or the person he was talking about above).
In this case, "Duncan the kid" is used for a specific purpose: so we know in which
direction the car is moving (the lines behind the car also help in this regard, as will
be seen below). Other children had interview responses similar to Duncan's, and
saw an explicit need to include some form of a reference so that it would be possible to communicate and make explicit the direction in which the car was moving.

Children's representations of people show that breaking down children's representations into individual elements can only take the analysis so far. The elements that the children do include have specific meanings in the context of a
specific goal, in this case representing the movement of a car. The next section
looks specifically at the tools the children use to represent movement.

Elements Children Use to Communicate Movement

Use of Smoke and Dust. Movement, unlike cars and people, is not a
physical object. However, some children brought physical elements into their representations to help them convey motion. One of these strategies was the use
of representations of smoke and dust. Six of the children's representations (8%)
included markings resembling smoke or dust. Children specifically identified the

lines, as in Figure 3.7, as smoke or dust (smoke in this specific case). Smoke always, regardless of the student, exists behind the car, opposite the direction of movement (as in, the car moves away from the smoke).

Use of Other Figurative Objects. Smoke and dust were not the only figurative elements used to show movement that the children brought into their drawings. Three representations contained figurative objects other than smoke or dust, and two are discussed in this section. For instance, Figure 3.8 shows a stop sign. Anthony explained that the drawing shows a car that does two actions in succession, moving quickly and then stopping. In the example in Figure 3.9, a car moves closer, farther away, and then stops. The stopping is shown by the person on the right of the picture holding a key, and Daniel described that the car cannot go anywhere if the key is not in it. The strategy of including a key was unique to this child, but it did show that some children needed to include an explicit cue about the (non)movement of the car.

Use of Lines. Another element that the children used was lines behind the car. Seventeen of the 76 representations (22%) included lines. Unlike smoke, dust,

Figure 3.7. James's Use of Smoke Behind the Car (on the Left-hand Side)

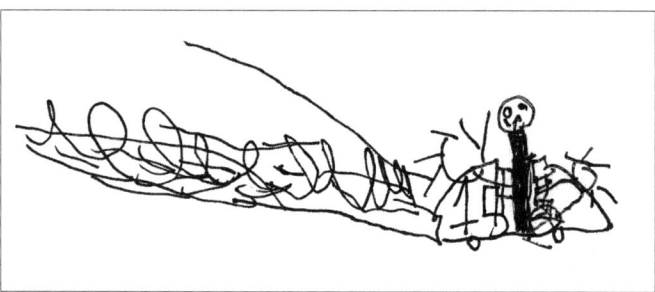

**Figure 3.8. Anthony Included a Stop Sign
(on the left of the Representation)**

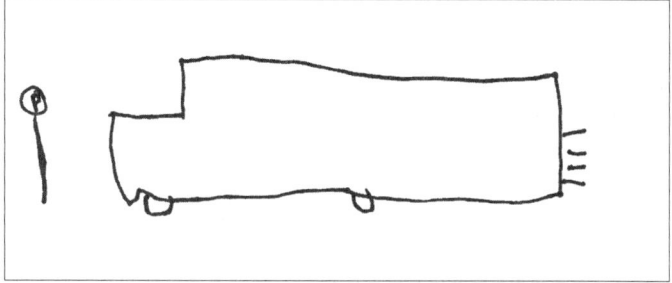

Figure 3.9. Daniel Uses a Held Key by a Person in the Drawing to Show a Stopped Car

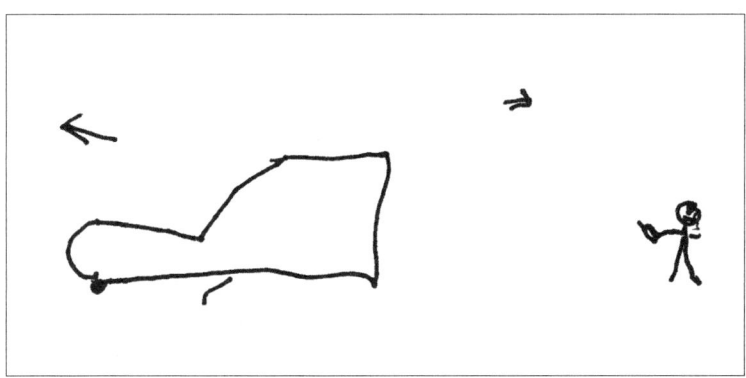

stop signs, or keys, lines are not something that the children would typically see in the physical world (this is not intended to discount that this approach for conveying movement can be common in other media such as cartoons or games; however, lines are not tangible, unlike the features mentioned above); however, Sherin (2000) did note that line segments are a constructive resource for representing movement (at least among older children). This is to say that line segments are a useful tool that children have at the ready for drawn representations. The lines used by students in this study had different forms. The most common was a line behind the car, such as in Figures 3.6, 3.7, and 3.8. Another less common strategy was to put lines completely around the car, as in Figure 3.2.

Use of Arrows. Arrows are similar to lines in the sense that they are non-physical entities that are brought into the representation. Fifteen representations (20%) included arrows. Lines typically were shown behind the car, but arrows were positioned either above the car or in front of the direction of movement of the car. Arrows typically pointed left or right, but two of the children created representations where the arrows pointed up (as in Figure 3.10) or down. Both up-and-down pointing arrows referred to the car being stopped.

Use of Writing. The last, somewhat commonly used tool in the representations was writing; seven representations (9%) included writing. Writing accomplished two purposes. Duncan used it to label the actors in the representation (as in Figure 3.6). Slightly more common was the use of writing to describe speed. In Figure 3.11, the words "*LOING FAT*" appear ("going fast"). In Figure 3.12, the child wrote "*SLOW*" above the car to show the car moving slowly. Additionally, one representation contained the word "*SDLE*" ("still"), to show a car not moving.

Figure 3.10. An Example of An Arrow Pointing Up to Show a Car Stopped (After Moving Closer and Farther, Shown by the Horizontal Arrows)

Figure 3.11. William Writes *LOING FAT* (Invented Spelling for "Going Fast") to Describe the Movement of the Car

Figure 3.12. An Example of Charles Using Writing to Show the Car Moving Slowly

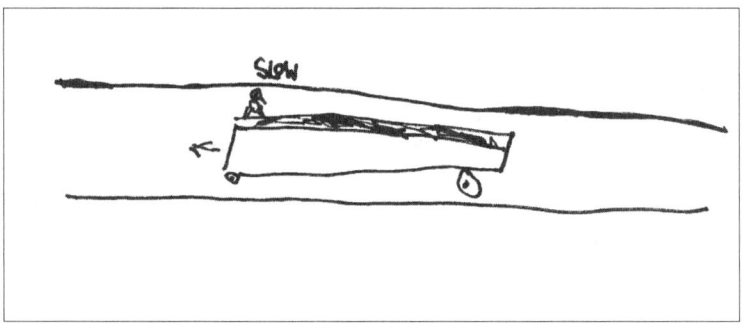

Representations of Motion Concepts

Thus far, we have looked at the tangible elements that the children have put on paper: cars, people, and markings on the paper. These elements comprise a tool set (Sherin, 2000) that the children used in making their representations of movement. Fifty-eight of the 76 representations (76%) contained some amount of explicit information about the car's movement, indicating an in-drawing assembly of these tools to create meaning about movement. This information related to either the speed or direction of the car. This section looks at how children manipulated their tools in order to build meaning about movement.

Representing Speed—Fast Versus Slow. Lines were one way that children used to show movement of a car. Lines were included in 17 representations, or 29% of the representations that explicitly communicated movement. In order to show speed, children typically added more lines. This gives us some idea about how these children conceive of speed; going faster is indicative of something more or bigger. Figure 3.13 is an example of how Nora added more lines to show a car moving faster. In this case, the lines always surrounded the car. When the representation showed the car going faster, she added a more dramatic, bolder, and bigger style of lines to show more emphasis to the car. It is as if in gaining speed, the car gained emphasis in the representation. Figure 3.14 shows a similar type of manipulation of a representation for showing speed. Anthony in this case differentiates between showing a slow-moving car and a fast-moving car by adding more lines behind the car. This strategy is also used with arrows; the third panel of Figure 3.11 shows two arrows next to the car. The child explained that this is to show that the car is moving faster.

Figure 3.13. Adding More Lines is Used to Show More Speed

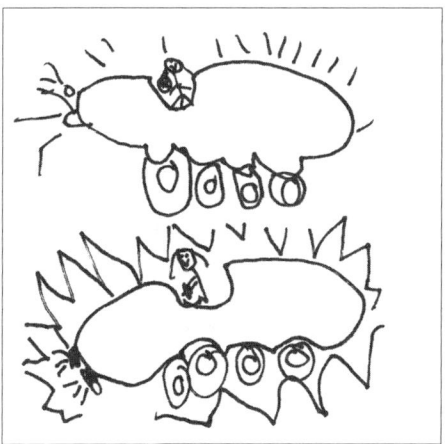

Figure 3.14. An Example of Anthony Adding More Lines Behind the Car to Show More Speed

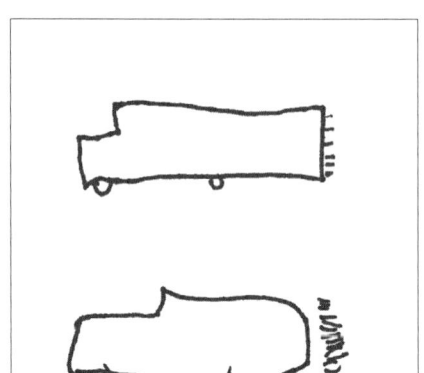

Another technique for showing speed was to manipulate the car's location within the physical space of the representation. In Figure 3.15, Amelia has drawn three successive pictures to show the movement of the car. In the first scene (at the top) the car has moved away from her. In the second scene, the car moves closer to her, quickly. She explained that one could tell this because the car has covered a lot of space on the piece of paper. In the last scene, at the bottom, the car has moved slightly farther away from her. She described that this is because the car has moved slowly away from her. Once again, this shows a little bit about Amelia's underlying concept of speed, in that she understands that things that move faster cover more distance. There is no evidence in the data about her thinking about time; in that, we do not know if she perceived the same amount of time passing between each scene. However, we can speculate that she has some implicit mechanism for considering time, as her drawings seem to presume a constant change in time across the three frames.

Children's thinking about speed as relative changes in movement was not completely isolated. Figure 3.16 is an example of William doing a similar thing. In this case, there are not three repeated scenes; instead, he has drawn the same car over and over again at different times. In between the car on the left and the car at the center, he described that because the car has moved quickly, there is a lot of space between the cars. In the subsequent movement, he drew less space between the middle and right cars. This is to show the car moving slower. The use of space in drawings of movement is well documented as a resource that children will use (Sherin, 2000). However, these children are using space in a sophisticated way. The amount of space that the car *moves* is relative to space elsewhere in the drawing. They are showing an understanding of speed as the distance something traverses, presumably in a fixed amount of time.

Figure 3.15. Showing Speed by Manipulating the Car's Physical Position in Relation to a Referent Within the Representation

Figure 3.16. William Showing Changes in Speed by Manipulating How Far the Car Moves from One Drawing to the Next

Representing Speed—Movement Versus Staying Still. Another type of speed is no movement. Among the children, there were many methods to show no movement. Duncan saw the lack of movement as an absence, and produced a representation that correlated with that understanding. Upon producing a representation of a car with no lines trailing it, he was asked about it. The child replied, "That means it's not moving." Seeing no speed as an absence, however, was not common, and in fact several children saw the need to add to their drawing to show no speed. Figures 3.8 and 3.9 show this. In the former, Anthony includes a stop sign and in the latter Daniel includes a key. Perhaps the most extreme example of this is shown in Figure 3.17. In this case, Nora drew a picture of a flower to show a car not moving. She explained that she chose to do this because a flower does not move.

Figure 3.17. A Flower Drawn by Nora to Represent a Car Not Moving

Representing Direction

As discussed before, 30 representations (40%) included a person in the drawing who was not connected to the car. When children were asked about the inclusion of people in their drawings, a common response was that the person was included to help make sense of what direction the car was moving, and whether the car was moving closer to or farther from a person. One kindergartner, Charles, summarized it succinctly, saying he included the person because "you have to know if it [is] going towards you or away." Figure 3.9 is an example of a child using a person to help show direction.

While there are frequently cues to show what direction the car is going, several children leave the idea of closer versus farther less explicit in their representation. This is particularly apparent in cases where children do not include a person in their drawings. However, even children who did not include a person in their representation still included cues about the direction of the car. The most common way of doing this was to either rotate or reflect some asymmetric element of the representation, either the lines behind the car, an arrow, or the car itself. For example, consider James's drawing in Figure 3.4, which includes a steering wheel and dust behind the car to give the car orientation. When he drew the car going in the other direction, he simply mirrored the elements that serve to orient the car.

When asking children about their representations without an explicit origin, a common question from the interviewer was "Where are you [in the drawing]?"

Figure 3.18. An Example Where Robert Includes Overlapping Representations for Motion Closer and Farther; the Same Car Contains Lines for Both Directions, but with no Indication of Sequencing.

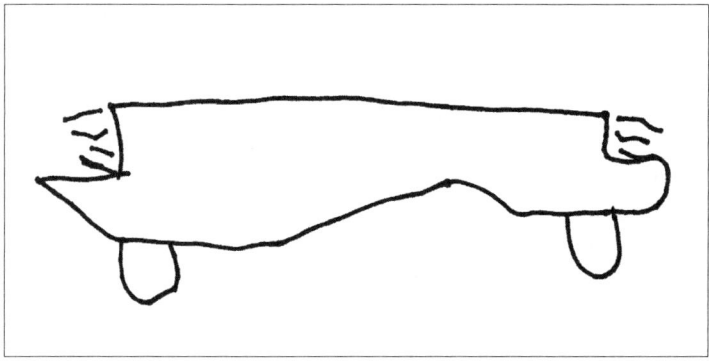

The question did not presuppose that the child was in the representation, and was used as a starting point to gauge if and how the children used themselves as a point of reference in the drawing. One response, from Duncan, who did not include himself in the drawing, described the section of the classroom that he was in, "the problem-solving loft." Soon after, when asked to describe the movement of the car in his drawing, he said that the car was moving "to the bathroom," another physical anchor within his classroom. These results hint that he sees his drawing as an extension of his environment. The car he drew in his drawing still moves in relation to the larger space. The car can move away from him, but it is worth keeping in mind that he does not include himself in the representation and, when asked to define where he is, in the context of a represented moving car, he responds by defining his location in the classroom. Further, some other children, in response to the question "Where are you [in the drawing]?" responded with "here," indicating that they were possibly thinking of the drawing as inextricably connected to themselves, a conjecture that needs to be followed up in future research.

Representing Change of Movement

The children in the study produced 31 representations of a car changing directions or speeds in the course of the movement. In 16 of these cases (51%), the children produced and described some system within their representation for differentiating these changes. One basic strategy was to re-create the scene multiple times, with the car and reference (if originally included) redrawn each time. This strategy is akin to comic book panels. Figure 3.4 is an example of a comic book style approach. Five of the 31 representations (16%) used this approach.

Another approach that the children took was to draw the moving car multiple times, but to change some aspect of the representation in a purposeful way to indicate that the car has changed its movement. Figure 3.16 is an example of this. In this drawing, the child drew the car repeatedly. However, in each drawing he changed the spacing between the cars to show how the car has changed its speed.

The final strategy observed was one where children used no repetition, but instead modified the car without any drawn indication of ordering. Figure 3.18 is an example of this, as are Figures 3.9 and 3.10. In these cases, multiple representations for directions are overlapped. The children provide no explicit marker for the sequence in which the movement unfolds. Sequencing is an area of developmental growth in children of this age (Piaget, 1929/2007). In cases where the child provided little explicit information about changes in movement, it may be the case that they are still developing ideas of sequence and narrative.

CHILDREN'S REPRESENTATION OF MOVEMENT IN CONTEXT WITH PREVIOUS WORK

Perhaps the most salient finding resulting from the research described here is that kindergartners' drawings of movement are coherent and systematic. The children may not have representational systems (e.g., notational systems) at the ready, but when prompted in the context of the task, the children are able to present a system that does not reflect a random selection of elements. Karmiloff-Smith (1990) questioned the systematicity of young children's drawings. However, by looking at how children use the same elements consistently, we can see that at least in the context of representing movement, these children were deliberate. Representations used a single, consistent vocabulary in communicating meaning, as representations do not include combinations of movement lines and arrows, but rather one particular element or the other. They use the space of the representation consistently, placing the movement lines in the same location in relation to the car, or using the car's location within the representation in predictable locations depending on the phenomenon they are trying to represent.

The children certainly have tools in their representational arsenal to think about movement. They can do so in a personally meaningful way, and create structure within their drawings that helps them communicate a narrative for movement. One could easily consider weaknesses in their approaches to drawing: The drawing system lends only to coarse distinctions of speed (fast versus slow, though the line approach could easily be extended in a systematic way), and for most children a point of reference is an implicit construct. However, it is equally important to examine the strengths of their representations. The children's drawings show that when asked to think about movement they have both conceptual and representational resources to do so. Two related ideas can help shape the children's ability. The first is epistemological resources (Hammer, 2000), which describe the meta-cognitive strategies learners use to extend their thinking. One of Hammer's concrete examples is the ability to draw in related ideas, and we see the children building their representations from their previous experiences. This is particularly apparent when we think about the elements that the children added to their representation but were not presented to them, like a stop sign or a key, both to show a car standing still. More directly, the idea of meta-representational competence (diSessa & Sherin, 2000; Sherin, 2000) is relevant to the data presented in this chapter. These children build consistent representations based on their previous experiences working with other representations (e.g., drawings) and their interactions with physical concepts of movement. They consider gross qualitative differences in speed (fast versus slow) and methods for showing no speed. The children consider problems of direction and some of them explicitly think about a point of reference. The children show an ability to manipulate their representations systematically by altering elements in their representations. These elements are lines or arrows, as well as the location of the car on the page. Lines, arrows, and mechanisms for sequencing are unseen; the representational elements do not have a tangible analogue in the physical world. The children make use of physical elements, as well as sometimes utilizing elements unseen, and build systems to create a motion narrative when the car changes directions. These are all skills that the child can draw and build upon when given opportunities to do so.

Conceptions of Movement

There are many features of the children's representations that could reveal more about their thinking about movement, but here I want to further examine the children's representations of speed and their use of explicit origins. When the children distinguished between fast and slow, they frequently added markers to show that the car is moving faster. This typically took the form of more lines or arrows. DiSessa (1988) suggested the idea of phenomenological primitives (p-prims) in describing scientific learning and development. P-prims are basic building blocks for more complex ideas to be formed, and are contextually used as children confront novel situations or form new understandings of familiar ones. One of the p-prims that diSessa proposed is "more is more," and it certainly seems that when it comes to showing more speed, many children show more speed by adding more emphasis to the representation. More lines, more arrows, and more emphasis show this across many of the representations. This shows us something about the children's conception of speed: It is quantity that can be added to, and they show this increase by adding to their drawings. Their representations of increases in speed are consistent with a "more is more" type of approach.

This line of thinking, however, presents a challenge when considering how children show no speed. If they consider no speed to be an absence, then a drawing with no lines or emphasis would convey the idea. While children's drawings of no speed tended not to include lines, the children often included a novel object; examples include a stop sign, a key, and a flower. Their representations of no movement reveal something about their conceptions of no movement, namely that no movement is qualitatively different from a moving object. The conventional view of no movement is a complete absence of speed. Children also have difficulty initially building a concept for zero (Hughes, 1986; Wellman & Miller, 1986), and while the same research has shown that the concept of zero finally takes root around the age of 4, perhaps it should not be surprising that kindergartners will not readily reach for an idea that is new to them and extend it to their representations of movement. In my work with kindergartners, they treated the idea of no movement as a qualitatively different scenario, where they added new elements into their drawings in order to communicate an absence.

The other striking element in the representations is how infrequently the children used an explicit origin in their representations. Pozo (1999) discussed the implicit and explicit idea of representation, describing that a representation is a chance to explicitly frame and consider certain ideas, but there may be implicit ideas within an individual's internal representations. While Pozo also presented a mechanism of change with implicit elements becoming more explicit with time and experience, it is useful to look at what elements the children leave implicit. When children exclude an origin, it is impossible for a third party to distinguish direction in the movement, such as moving closer versus moving farther way. It seems unlikely that the children do not have a framework for closer and farther away. Rather, it seems more likely that when the children do not include a point of reference, the reference is implicit. A plausible explanation is that children find an explicit representation of the origin (or point of reference) unnecessary because the representation is an extension of themselves; the car moves in relation to their physical self, and there is no need to re-represent themselves. If including an explicit point of reference is a developmental trajectory, then it seems likely that older children would more readily include a reference. In Sherin's (2000) research, most of the children's drawings included a cactus, which was a fixed object in the motion narrative he presented (but representing the self as referent versus an outside object as referent could likely pose additional challenges for children).

CONCLUSION

Children have resources to build representations of motion, and these representations are useful in exploring children's developing understanding of movement. As we learn more about young children's representational skills, both in movement and elsewhere, I suspect we will continue to find a wealth of abilities, and be able to use this knowledge to further enrich children's exploration of movement and other scientific phenomena.

ACKNOWLEDGMENTS

I am greatly indebted to Bárbara M. Brizuela and Ronald Thornton, my dissertation advisors. Their intellectual fingerprints can be found throughout this paper.

REFERENCES

Brizuela, B. (1997). Inventions and conventions: A story about capital numbers. *For the Learning of Mathematics, 17*(1), 2–6.

Danish, J. A., & Enyedy, N. (2007). Negotiated representational mediators: How young children decide what to include in their science representations. *Science Education, 91*(1), 1–35.

de Saussure, F. (1931). *Course in general linguistics.* New York: McGraw-Hill.

diSessa, A. A. (1988). Knowledge in pieces. In G. Forman, & P. Pufall (Eds.), *Constructivism in the computer age* (pp. 70–79). Hillsdale, NJ: Lawrence Erlbaum Associates.

diSessa, A. A., Hammer, D., Sherin, B., & Kolpakokski, T. (1991). Inventing graphing: Meta-representational expertise in children. *Journal of Mathematical Behavior, 10,* 117–160.

diSessa, A. A., & Sherin, B. L. (2000). Meta-representation: An introduction. *Journal of Mathematical Behavior, 19*(4), 385–398.

Duckworth, E. R. (1973). The having of wonderful ideas. In M. Schwebel & J. Raph (Eds.), *Piaget in the classroom* (pp. 258–277). New York: Basic Books.

Gardner, H. (1982). *Artful scribbles: The significance of children's drawings.* New York: Basic Books.

Ginsburg, H. P. (1997). *Entering the child's mind: The clinical interview in psychological research and practice.* New York: Cambridge University Press.

Goodman, N. (1976). *Languages of art: An approach to a theory of symbols.* Indianapolis, IN: Bobbs-Merrill Co.

Hammer, D. (2000). Student resources for learning introductory physics. *American Journal of Physics, 68*(S1), S52–S59.

Hughes, M. (1986). *Children and number.* Cambridge, MA: Blackwell.

Karmiloff-Smith, A. (1990). Constraints on representational change: Evidence from children's drawing. *Cognition, 34*(1), 57–83.

Martí, E., & Pozo, J. I. (2000). Más allá de las representaciones mentales: La adquisición de los sistemas externos de representación. *Infancia y Aprendizaje, 90,* 1–4.

McDermott, L. C., Rosenquist, M. L., & van Zee, E. H. (1987). Student difficulties in connecting graphs and physics: Examples from kinematics. *American Journal of Physics, 55*(6), 503–513.

Nemirovsky, R., & Ferrara, F. (2009). Mathematical imagination and embodied cognition. *Educational Studies in Mathematics, 70*(2), 159–174.

Nemirovsky, R., & Noble, T. (1997). On mathematical visualization and the place where we live. *Educational Studies in Mathematics, 33*(2), 99–131.

Ochs, E., Gonzalez, P., & Jacoby, S. (1996). "When I come down I'm in the domain state": Grammar and graphic representation in the interpretive activity of physicists. In E. Ochs, E. E. Schegloff, & S. A. Thompson (Eds.), *Interaction and grammar* (pp. 328–369). Cambridge, UK: Cambridge University Press.

Piaget, J. (1970). *The child's conception of movement and speed.* New York: Basic Books.

Piaget, J. (2007). [1929]. *The child's conception of the world.* Lanham, MD: Rowman & Littlefield Publishers.

Pozo, J. I. (1999). Más allá del cambio conceptual: El aprendizaje de la ciencia como cambio representacional. *Enseñanza de las ciencias, 17*(3), 513–520.

Sherin, B. (2000). How students invent representations of motion: A genetic account. *Journal of Mathematical Behavior, 19*(4), 399–441.

Wellman, H. M., & Miller, K. F. (1986). Thinking about nothing: Development of concepts of zero. *British Journal of Developmental Psychology, 4*(1), 31–42.

Young Children's Appropriation of Representations

Beth Warren and

Ann S. Rosebery

The chapters in this section, *Appropriation of Representations*, highlight investigations into young children's use of representations in mathematical, scientific, and spatial domains. In these studies, "young" refers to children attending preschool and Kindergarten, ranging in age from just shy of 3 to 6.5 years old. The chapters address a range of representational forms (e.g., dots on dice, invented drawings, self-constructed maps) and a varied conceptual landscape (e.g., discrete quantities, movement, and maps of local routes). As varied as the chapters are, they are also connected to a common concern with understanding children's use, invention, and production of representations for symbolic purposes. In this essay, we begin by briefly discussing this common focus. We then offer some thoughts that arose as we considered the chapters through the prism of our research and recent work by the anthropologist Tim Ingold (2011).

These chapters explore the conceptual and representational resources that children demonstrate in the context of varied types of production tasks involving symbol use. Martí, Scheuer, and de la Cruz posit that using representations symbolically involves understanding their "double nature: They are physical objects with a given form, color, texture, and position in space, yet at the same time they represent properties that refer to, and therefore can be transferred to, other situations." Each of the studies in this section reports evidence to support the view that young children demonstrate varied understandings and engagements with the symbolic nature of representations.

Kahn, for example, shows that the kindergartners he interviewed used "nonphysical" elements such as lines and arrows systematically to represent aspects of speed (e.g., by varying the number, length, or intensity of lines to show more or less speed) when asked to draw the movement of a toy car. Further, he found that a number of the children used space within their drawings to convey meaning about the car's movement, i.e., the amount of space that a car is shown moving is relative to space elsewhere in the drawing. Kahn infers that in their use of space these children demonstrated an understanding of speed as the distance something traverses, presumably in a fixed amount of time.

Brizuela and Cayton-Hodges found that the route maps made by kindergartners changed from being "drawing-like" (e.g., inventories of objects seen on their routes) to being more "map-like" (e.g., schematic use of arrows and lines to show the route, with landmarks noted along the way). They also found that in their later maps some children incorporated more conventional notations (e.g., crosses and dots to mark starting and ending points of routes).

Martí, Scheuer, and de la Cruz used a game-like task to investigate whether preschool-aged children understood correspondences between the number of dots on the face of a die and the number of spaces a toy horse should be moved on a game board. They found that the children engaged with the quantitative symbols (i.e., the number of dots on the face of the die) and the rules of the game (i.e., the number of spaces the toy horse should be moved) in varied ways, ranging from children who did not seem to understand the correspondences to those who showed full mastery.

These studies contribute to the field's understanding of the constructive resources that young children bring to bear on understanding, using, and creating representations for symbolic purposes in complex scientific and mathematical domains. While not an explicit focus, the findings raise for us the question of *where, when, and how* children learn to create and use representations symbolically. Intriguingly, from our perspective, each study hints at some possible ways of thinking about this question.

Brizuela and Cayton-Hodges compared kindergartners' self-constructed maps at the beginning and end of a school year. Throughout the year, as a supplement to the regular curriculum, the children engaged in a series of activities designed to introduce them to various aspects of maps and map-making (e.g., measurement and scale, map interpretation, and map production). Brizuela and Cayton's study, particularly their discussion of the maps created by Callie, suggest to us that map construction may be understood as part of children's dynamically unfolding, lived, meaning-making experiences in the world.

Martí, Scheuer, and de la Cruz report that some of the children in their study (those in Group 3) made "progress in their responses" as they proceeded through different sections of the game-like task involving discrete quantities. They note that the responses of these children showed greater integration of the rules of the game with the quantitative information (i.e., the correspondences between the number of spaces the toy horse should be moved and the number of dots on the face of the die). Although only limited detail of the children's engagement with the task is provided, we are intrigued by the suggestion that these 3-year-old children adapted their responses *as they played* the game.

Finally, while the goal of Kahn's study was to explore children's thinking about movement, he views the conceptual and representational resources they deployed in their drawings as evidence in part of their worldly experiences with movement. He addresses this with regard to the children's use of lines. He notes that the children may likely encounter lines as a way to symbolize movement in cartoons or games or other media—especially, we might add, in a world saturated by digital media.

Taken together, along with our own research, these suggestions make sense to us: Children develop rich repertoires of meaning-making practices as they experience and draw movement, walk and map routes, and encounter discrete quantities in myriad ways in their everyday lives (Gutiérrez & Rogoff, 2003; Nasir, Rosebery, Warrren, & Lee, 2006). In an exploratory classroom-based study, we observed 1st- and 2nd-graders keenly attuning their attention to various elements used in children's books to convey changes in speed, including lines, color, and reference points (e.g., a lamppost repeated on successive pages to show changes in the position of a vehicle or person). The children then creatively appropriated these elements in their own drawings of movement as well as in constructive critiques of one another's drawings, specifically with respect to symbolizing change in speed (see Rosebery, Ogonowski, DiSchino, & Warren, 2010; Warren, Ogonowski, & Pothier, 2005, for discussion of related studies). The children's work in this study inclines us toward seeing physical enactments, analysis of storybook conventions, and drawings as nonseparable and nonhierarchical aspects of their imaginative activity. As Tim Ingold (2011) argues, "the terrains of the imagination and the physical environment run into one another to the extent of being barely distinguishable" (p. 198).

Building from Ingold's perspective, we would like to offer readings of two of the representations discussed in this section's chapters. First, from Kahn's study, we consider Nora's striking drawing (see Figure 3.2 in Chapter 3): a fanciful car, with a broadly smiling driver, surrounded by dynamically radiating lines. To our way of looking, Nora's drawing evokes the felt joy of speed experienced in a cool car. In her drawing, the mental and the material, the symbolic and the physical, are not bordered territories. Note, for instance, how her drawing nearly bursts the edges of the picture frame; the felt experience of speed cannot be contained within the boundary of representation.

For the second example, we turn to the first home-to-school route map produced by the student Callie in Brizuela and Cayton-Hodges's study (see Figure 2.13 in Chapter 2). Unlike Callie's first map of the school-to-park route (see Figure 2.11 in Chapter 2), this map traces a route with starting and ending points explicitly marked. Brizuela and Cayton-Hodges speculate that this more "schematic" route might reflect Callie's daily experience of a 30-minute bus ride to and from school. Perhaps, rather than being more schematic than her rendering of the less familiar school-to-park route, this map *reveals aspects of her experience of movement* on (and with) the bus. Perhaps her image is as much related to her kinesthetic participation in the journey as to her noticing elements such as land-marks along the way. Taking this view, her later map of the same route (see Figure 2.14 in Chapter 2) suggests to us that she is *experiencing* the bus trip differently in June from the way she did in September; she is attuned to and attentively engaged with different aspects of the journey (Ingold, 2001).

Our reading of these selected examples as well as our own classroom-based research inclines us strongly toward the view, articulated by Ingold (2011), that observation, participation, and description are not separable: "To move, to know, and to describe are not separate operations that follow one another in series, but rather parallel facets of the same process—that of life itself" (Ingold, 2011, p. xii). Working from this integrated, nonhierarchical perspective on what it means to "represent symbolically" may help further widen the lenses used in developmental and learning sciences research to understand the meaning-making resources that young children, viewed as symbol-users (Nemirovsky, 2005), dynamically develop through their engaged participation in the varieties of everyday life.

REFERENCES

Gutiérrez, K. D., & Rogoff, B. (2003). Cultural ways of learning: Individual traits or repertoires of practice. *Educational Researcher, 32*(5), 19–25.

Ingold, T. (2001). From the transmission of representations to the education of attention. In H. Whitehouse (Ed.), *The debated mind: Evolutionary psychology versus ethnography* (pp. 113–153). Oxford, UK: Berg.

Ingold, T. (2011). *Being alive: Essays on movement, knowledge, and description.* London and New York: Routledge.

Nasir, N., Rosebery, A., Warren, B., & Lee. C. D. (2006). Learning as a cultural process: Achieving equity through diversity. In K. Sawyer (Ed.), *The Cambridge handbook of the learning sciences* (pp. 489–504). Cambridge, UK: Cambridge University Press.

Nemirovsky, R. (2005). Mathematical places. In R. Nemirovsky, A. Rosebery, J. Solomon, & B. Warren (Eds.), *Everyday matters in science and mathematics: Studies of complex classroom events* (pp. 45–94). Mahwah, NJ: Lawrence Erlbaum Associates.

Rosebery, A., Ogonowski, M., DiSchino, M., & Warren, B. (2010). "The coat traps all your body heat": Heterogeneity as fundamental to learning. *Journal of the Learning Sciences, 19*(3), 322–357.

Warren, B., Ogonowski, M., & Pothier, S. (2005). "Everyday" and "scientific": Re-thinking dichotomies in modes of thinking in science learning. In R. Nemirovsky, A. Rosebery, J. Solomon, & B. Warren (Eds.), *Everyday matters in science and mathematics: Studies of complex classroom events* (pp. 119–148). Mahwah, NJ: Lawrence Erlbaum Associates.

Part II

MAKING MEANING

CHAPTER 4

Students' Handling of Graphs at the University Level

*María-Puy Pérez-Echeverría, Yolanda Postigo,
and Cristina Marín-Oller*

The general aim of our work is to analyze the way in which social science students read, interpret, produce, and use graphs with the implicit aim of being able to contribute to better teaching of these aspects. Within this field, the specific goal of the study we describe below was to investigate which factors have an influence on students' interpretation of graphs by means of a task whose objective is to relate data to different graphs.

Data graphs are one of the most frequently used external representation systems within social sciences. According to the evolutionary view of the development of the human mind presented by Donald (1991), representation systems that evolved after language became cultural artifacts necessary for the development of the theoretical mind. In our opinion, current external representation systems fulfill a function similar to that set out by Donald. In other words, these cultural systems are true thought amplifiers (Lévi-Strauss, 1963; Martí, 2003; Martí & Pozo, 2001; Olson, 1994) that contribute to the development of an expert or specialized mind, capable of interpreting or creating new meanings in the sense that they enable both communicative and epistemic functions. In this context, they "format" the minds of those designing, interpreting, or using them, restricting and giving meaning to different kinds of information (Olson, 1994; Pérez-Echeverría & Scheuer, 2009). Therefore, it is not surprising that different fields of knowledge can be characterized, at least in part, by the development of external representation systems that enable both the development of specific meanings and an efficient communication. Nonetheless, these functions are not evident. The specialization and specific development of these systems make most of the relationships among and within systems and their functions not obvious, which leads to difficulties in their use and in their learning and teaching (Pérez-Echeverría, Martí, & Pozo, 2010; Pérez-Echeverría & Scheuer, 2009).

EXTERNAL REPRESENTATION SYSTEMS IN THE SOCIAL SCIENCES

As we stated previously, within social sciences, data graphs are one of the most frequently used external representation systems (Butler, 1993; Peden & Hausmann, 2000), as they are a key tool to communicate and understand data that reveal the usually complex relationships among variables in research, and they also provide models that allow the development of theoretical proposals (Latour, 1990; Roth & Bowen, 2003). Therefore, we can infer that the ability to read, interpret, and construct graphs should form part of the set of abilities to be developed by social scientists during their formal education, regardless of the professional field in which they will practice in the future.

The study we discuss in this chapter forms part of a broader research project aimed at studying which graphical comprehension skills social science students have developed, especially within the field of psychology, during their education (Pérez-Echeverría, Pecharromán, & Postigo, 2007; Pérez-Echeverría, Postigo, & Marín, 2010; Pérez-Echeverría, Postigo, & Pecharromán, 2009; Postigo, Pérez-Echeverría, & Marín, 2010). In other words, our goal is to analyze students' "graph sense" (Friel, Curcio, & Bright, 2001) in these fields, and investigate which variables have an influence or determine how these students interpret graphs. Specifically, in this chapter we will present data about how social science students (specifically, in psychology and economics) at the college level choose which they think is the most suitable graph for the presentation of results from different research studies.

Teaching and Learning Graphs

Studies on experts reveal that proficiency is specifically associated with the skills to quickly, and in an overall way, interpret graphs that present information from their specialty, and to quickly and precisely draw inferences on the content of these graphs (Roth & Bowen, 2003). Nonetheless, there are barely any studies that reveal how this skill has developed or how formal instruction can contribute to the acquisition of this skill, once the rudiments of graphical language have been learned or, in the words of Balchin and Coleman (1965), "graphicacy" has been acquired. As is true in other countries as well (see, for example, Romberg, Fennema, & Carpenter, 1993), the initial teaching of graphs in Spain takes place during the primary school (among 6- to 12-year-olds) mathematics curriculum and also during the compulsory secondary education curriculum (among 12- to 16-year-olds). During the final years of primary school, students are initially introduced to graphs and other representations, especially tables. The main goal of this initial introduction is for students to develop the ability to organize and condense sufficient information to interpret and construct simple tables that enable them to solve certain problems effectively, although this result is not always achieved (see some difficulties with this kind of representation in Gabucio, Martí, Enfedaque, Gilabert, & Konstantinidou, 2010; Martí, Gabucio, Enfedaque, & Gilabert, 2010). In addition, students are also expected to learn to read graphs of very simple data.

Later on, during compulsory secondary education, graphical representations are approached as expressions of mathematical functions or representations of statistical data. Thus, at this time teaching of graphical representations is normally focused on the translation of different algebraic expressions or tables of data into graphical formats or, conversely, the translation of graphs into algebraic expressions and tables. In these cases, the content and context of the tasks are treated as a mere accident or support so that mathematical processes can take place. Therefore, usually there is no instruction aimed at relating the graphical representations to different content (history, social content, and so on). Consequently, in spite of the fact that textbooks, especially in the social sciences, use graphs to report data or research studies backing up various theories, there is no teaching that emphasizes the role of these graphs to represent, communicate, and consider diverse information.

This way of teaching graphs exclusively from a mathematical point of view, with no relation to the specific knowledge or information they try to represent, complies with a rationalistic conception of curriculum (Eisner, 1974, 1994; Gimeno Sacristán, 1988), according to which it is sufficient to verbally teach or present certain content for it to be learned and applied to different situations and contexts. This kind of conception is especially strengthened by the characteristics of graphs themselves. Graphs are mixed representations that combine verbal, numerical, and iconic symbols (Martí, 2003). Underlying both this and other kinds of image-based external representation systems is the assumption about the simplicity of their interpretation (Pérez-Echeverría, Postigo, López Manjón, & Marín, 2009; Roth, Pozzer-Ardenghi, & Han, 2005). These assumptions consistently contribute to the idea that it is sufficient to teach syntactic aspects of graphs to learn syntactic, functional, and pragmatic aspects.

However, university social science students, including psychology and economics students, dedicate a large part of their efforts to topics related to statistical methodology and analysis of data. Usually, students studying these subjects use computer programs that automatically generate graphical representations of the information under study. In addition, the majority of students read scientific articles, books, or documents in which the results of research studies are reported both in graphical and textual formats. Nonetheless, only on rare occasions prior to university is work in the classroom aimed at specific analyses of graphical information or the relationship between graphs and texts. Therefore, although these students are immersed in an academic and cultural context in which graphs are very important, we cannot ensure that this context enables the development of skills and abilities needed by a social science expert.

Our own experience teaching psychology at the university level entails asking ourselves about students' abilities. An anecdote might serve as an example. Taking advantage of a typographical error in a research article that implied a clear contradiction between the graphical and textual information of the same data, we asked our students to interpret and explain the results of the study based on the graphical information and the conclusions of the authors. Students carried out this task without realizing the contradiction between the data provided in the graphs and in the text, and instead focused on the textual presentation of the data, ignoring the graphical information, which may not surprise us if we consider the predominance of verbal information in educational contexts.

Interpreting and Understanding Graphs

It is likely that the above anecdote is a consequence of the fact that understanding a graph in a way that is useful to assess a research study, to analyze theoretical content, or to infer the consequences of a certain model implies using very complex skills that involve the systematic combination of a broad set of aspects of knowledge (Kosslyn, 2006).

Some studies distinguish graphical knowledge, conceptual knowledge, and a set of contextual and personal factors that influence the graph reader's goals and interests (see, for instance, Shah, Freedman, & Vekiri, 2005). Nevertheless, it is difficult to discriminate between the effects of graphical knowledge and the effects of conceptual knowledge as, among virtually all experts, these two kinds of knowledge are related and it is difficult to separate the effects of one from the other. However, it is important to differentiate these types of knowledge to identify the different factors involved in their formal instruction (Postigo & Pozo, 2000; Stern, Aprea, & Ebner, 2003).

Many studies about graph comprehension (for a review, see Friel, Curcio, & Bright, 2001; Kosslyn, 2006; Pérez-Echeverría, Postigo, & Marín, 2010; Roth & Bowen, 2003; Shah, Freedman, & Vekiri, 2005) focus on the importance of perceptual characteristics of graphs as main factors that influence their reading and comprehension. Characteristics that deal with the elements, colors; or contrasts, the relative direction, size, or format of the representation may influence the reading of the data to a greater or lesser extent. Therefore, for example, most studies stress that in general it is simpler to interpret bar graphs whose interpretation enables one to make one-to-one comparisons of data, rather than continuous line graphs whose reading requires an overall interpretation and analysis of the information trends (Carswell, Emery, & Lonon, 1993).

In turn, the kind of graphical presentation one chooses depends on the kinds and complexity of the variables or data represented. Consequently, it is simpler to read discrete variables, in which the data may be interpreted in a more isolated way, rather than continuous variables. We took these studies into account when we designed the empirical tasks of the research described in this chapter. We assumed that our participants could choose a non-right graph (for instance, a bar graph for continuous variables) because this is easier to read and interpret. In addition, we also assumed that the more variables that are included in a graph and the more complex their interaction, the more difficult it would be to read and interpret the graph (Carpenter & Shah, 1998; Postigo & Pozo, 2000). All these aspects are related to the syntax of a graph. Like other external representation systems, it is possible to differentiate between syntactic and semantic aspects of a graph or, in other words, between the rules of the graph construction and its meanings.

Nonetheless, the ease or difficulty of interpretation depends not only on these factors related to the syntax of the representation, but also on how these syntactic aspects interact with people's conceptions about what is an appropriate graph (Pérez-Echeverría, Postigo, & Pecharromán, 2009), the graph readers' knowledge of the content depicted, the phenomena they are trying to explain, and their goals and interests when reading graphs (Friel et al., 2001; Roth & Bowen, 2003; Shah et al., 2005).

The graph reader's degree of knowledge about the content depicted in the graph also has an influence, together with the context or characteristics of the task itself, on the aims we establish and the interest with which we tackle the task. Consequently, the effort involved in different reading tasks and the consequences and inferences constructed by the reader are also very different (Gelman, Pasarica, & Dodhia, 2002; Meyer, Shinar, & Leiser, 1997). Therefore, our knowledge of the content represented in the graphs has an influence on the way we approach the task, but furthermore, the way in which we approach a task is also determined by our experiences as a graph reader (see for example Carpenter & Shah, 1998; Shah et al., 2005). Among the most apparent effects of this skill is the speed of reading and interpreting the graphs, the quick detection of inconsistencies or errors, and the relating of results with other data or theoretical knowledge.

As outlined above, the goal of the study we describe below was to investigate which factors have an influence on students' interpretation of graphs by means of a task whose objective is to relate data to different graphs.

HOW WE STUDIED THE SOCIAL SCIENCE STUDENTS' GRAPH SKILLS

The Tasks

To achieve this goal, students had to read four reports, each of them with five corresponding graphs, and to select the graph most suitable to the data presented in the report. We used experimental reports drawn from textbooks used to teach psychology in Spain. Together with the original text and the graph included by the textbook author, we included another four graphs and we asked psychology and economics students to indicate which graph they believed to be most appropriate to report the experimental data and the reasons for which they selected this graph. Therefore, these tasks required students to analyze whether or not the representation presented in each one of the graphs fit the textual information included in the textbook, and complied with the kinds of variables handled and the context. The different graphical options were designed to meet these aims and they were based on the studies about perceptual characteristics of graphs reported earlier.

The four reports were selected based on the fact that the complexity of the data and difficulty to interpret them were different. The content of the tasks had been part of the curriculum previously studied by the psychology students. For this reason, we assumed that the content was known by this group of students. On the other hand, economics students had more statistical instruction than their psychology peers.

We also designed another task with identical characteristics and identical graphical representations to portray data on studies about health instead of psychology. We assumed that the degree of knowledge about that content was similar among psychology and economics students. Therefore, this design allowed us to compare the effects of content and statistical instruction on participants' performance on tasks.

Each participant completed a questionnaire with four different research reports, which we call "problems." Each problem differed in terms of number, type, and kind of relationship among variables represented in the graphs. There were two different kinds of questionnaires, depending on their content: "psychological" and "nonpsychological." The "psychological content" questionnaire referred to content extracted literally from psychology textbooks, whereas the "nonpsychological content" questionnaire was about daily health problems, designed by the authors of this chapter (see Table 4.1 and the Appendix for an example of problem 2—psychological content).

Each problem consisted of a brief written research report extracted from a psychology textbook, where the aims and results of a study were explained. Below that information, we presented five different graphs, corresponding to the one used in the textbook and another four that showed different kinds of erroneous mistakes (see Appendix). Participants had to choose the graph that best matched

Table 4.1. Characteristics of the Problems

Problem	Variables	Psychological Content	Nonpsychological Content
1	One categorical variable	Scanning memory task	Effectiveness of therapies to stop smoking
2	One continuous variable	Effect of delayed reward on the speed of rats	Timetable effects of different slimming diets
3	Two continuous variables free of interaction	Effectiveness of extinction and omission training on behavior suppression in rats	Effects of anti-cholesterol treatments
4	Two continuous variables with interaction	Effects of induced learned helplessness in solving tasks in introvert and extrovert people	Kinds of obesity and effectiveness of different diets

the reported results. Mistake 1 implies that readers do not consider the characteristics of the variables. Mistake 2 refers to conceptions about the nature of the experiment (see mistake 2a in Table 4.2) or to textual aspects of the graph (axis label, legend; see mistake 2b). Although mistakes 2a and 2b are of a different nature, we consider them together because both imply a greater degree of misinterpretation regarding the problem than mistake 1 and less misinterpretation than mistakes 3 or 4. These last mistakes imply a misinterpretation of the data in the problem.

Who Are the Participants in This Study?

Seventy-five students took part in this research study: 38 economics students (6th semester) and 37 psychology students (8th semester), all of them from the same university in Spain. The groups differed in terms of their prior statistical knowledge. Economics students had already studied several subjects related to mathematics (two courses), descriptive and theoretical statistics (three courses), and econometrics (one course), while psychology students had studied statistics (two courses) and psychometrics (one course). Each of the two groups of students was in turn randomly divided into two groups according to the content of the questionnaire completed (psychological/nonpsychological [health]).

How Did We Analyze the Data?

We carried out a factorial study (2x2x4) with four experimental groups. We considered three independent variables: degree course (inter-subject with two levels: psychology and economics), content (inter-subject with two levels: psychological

Table 4.2. Characteristics of Each Graph

Graph	Characteristics
Correct graph	This graph is exactly the same as the one presented in the original textbook.
Mistake 1	Data matched the results explained in the text, but were not presented in the appropriate format: linear graphs (for categorical variables) and bar graphs (for continuous variables).
Mistake 2	Here two different cases were considered: 2a. In one-variable problems (1 and 2), this graph consists of a theoretical representation, showing the correct pattern but not the variability that usually appears when representing an experiment. 2b. In two-variable problems (3 and 4), the original graph was presented but labels for variable levels were reversed in the graph and the legend.
Mistake 3	Data presented in this graph showed a different pattern compared to the original one, but not contrary to it.
Mistake 4	Data presented in this graph were exactly opposite to that presented in the original one.

and nonpsychological), and difficulty of the problem (intra-subject with four levels: problem 1, 2, 3, and 4).

We analyzed the (1) influence of the content variable (psychological–nonpsychological) by means of the Mann-Whitney U test; (2) the difficulty of the four problems in the questionnaires in both content areas (Friedman's test and Wilcoxon's text); and (3) the kind of mistake (chi-squared test).

RESULTS

The analysis on the influence of the content variable (psychological and nonpsychological) reveals the same results among the psychology and economics students. We did not find statistically significant differences between the two kinds of content in any of the problems. Therefore, we report the analysis performed on the difficulty of the problems and the kinds of mistakes jointly for both content areas. For the presentation of results, we will first focus on the global performance in the task, based on the percentage of students who selected the correct graphs, to continue analyzing the difficulties of the tasks and the kinds of mistakes made by the students.

Selection of the Correct Graph

The results obtained by psychology students in the tasks proposed are presented in Figure 4.1, while the results from economics students are shown in Figure 4.2. As can be seen, neither of the groups shows a very high performance in the tasks. If we focus on the percentage of correct responses, less than 50% of psychology students chose this option for problems 2, 3, and 4. In problem 1, in which just one discrete variable was presented, the percentage of correct responses was 62%.

The percentage of correct responses by economics students is 65% for problem 1, while the percentage of correct responses is less than 31% for the other

Figure 4.1. Percentage of Correct Responses and Mistakes in the Four Problems (Psychology Students)

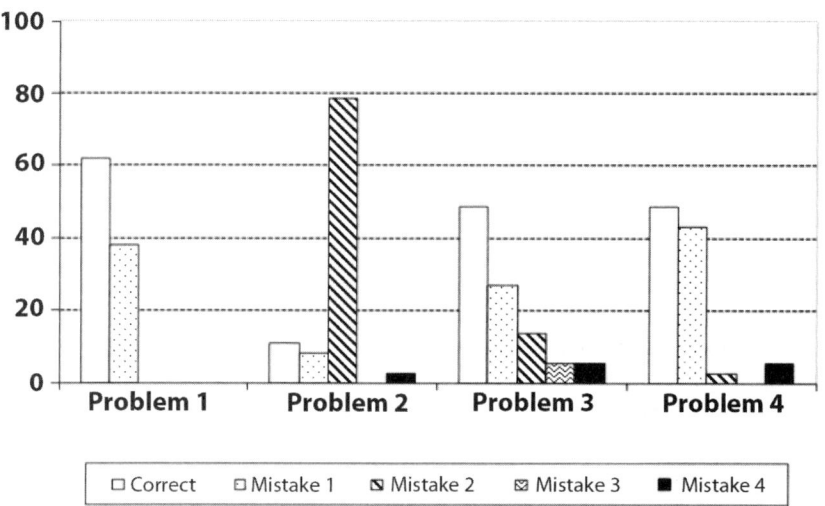

Figure 4.2. Percentage of Correct Responses and Mistakes in the Four Problems (Economics Students)

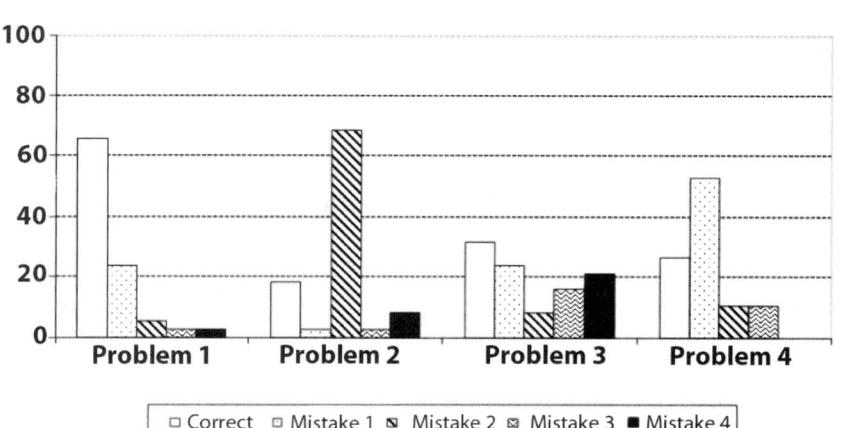

three problems (i.e., problems 2, 3, and 4). Therefore, in both student groups the performance is similar and lower than what we might expect among students with their training. Nonetheless, as we will assert in the conclusion, we cannot state that students had a completely inappropriate execution in the task, since the graphs that reflected most misinterpretations (mostly mistakes 3 and 4) were in general those less frequently selected by both groups.

What Kinds of Problems Are More Difficult?

The level of difficulty of the problems measured based on correct responses was similar across both student groups. Nevertheless, there are statistically significant differences between the four problems in both groups, as Friedman's test revealed. Therefore, we can differentiate several levels of difficulty in the problems based on the percentage of correct responses. In the case of psychology students, problem 1 had a higher percentage of correct responses than problems 2 and 3. Problems 3 and 4 had a higher percentage of correct responses than problem 2. There were no significant differences between problems 3 and 4.

In the case of economics students, problem 1 had a significantly higher percentage of correct responses than problems 2, 3, and 4. Problem 4 had a higher percentage of correct responses than problems 2 and 3.

Therefore, we see that for both groups of students the simplest problem was problem 1, while the most difficult problem was problem 2. This result is confirmed in the analysis in which we compared psychology and economics students by means of the Mann-Whitney U test for each one of the four problems. The test reveals statistically significant differences between the two groups exclusively in problems 3 and 4. It reveals a better performance among psychology students than economics students in problem 3 to the extent that they have a higher percentage of correct responses (48.6% compared with 31.6% among economics students) and fewer incorrect responses (mistakes 3 and 4). Only 5.4% of psychology students chose these mistakes, while 15.8% of economics students chose mistake 3 and 21.1% mistake 4. For problem 4, the differences between psychology and economics students also indicate a better performance among psychology students. In this case, the main difference between both groups is the percentage of correct responses among psychology students (48.6%) compared with economics students (26.3%).

What Graphs Are Misinterpreted?

The chi-squared tests for a sample analyzing the kinds of mistakes selected for each problem show statistically significant differences between psychology and economics students, indicating which kinds of misinterpretations vary based on the kind of problem.

For problem 1, the chi-squared test reveals that the differences in the graphs selected by economics students are statistically significant while this is not the case for psychology students. This is because in this latter case, all the selections, other than the correct response, are focused on mistake 1. This answer pattern, in which the most frequently selected graphs after the correct one were mistake 1 (e.g., selecting a line graph for categorical variables), only occurs among economics students for problem 1, while for psychology students this occurs in all the problems with the exception of problem 2. Mistake 1 was also the most frequent one chosen in problem 4 among economics students, although in this case mistake 1 was selected twice as frequently as the correct one (see Figure 4.2). For problem 3, there are no significant differences between the selections made by economics students because all the mistake graphs are chosen in similar proportions.

To sum up, this response pattern reveals that the most frequently selected response following the correct one was mistake 1, that is, those graphs that correctly represent the research data but in an inappropriate format (bars for continuous variables and lines for categorical variables). Therefore, although this selection reveals a lack of understanding of some syntactic aspects of graphs (the graphical format), it also indicates that the students are capable of pairing the research data reported in the text with a graph.

However, problem 2, the most difficult for students if we look at the number of correct responses, shows a different data pattern common to both groups. The distribution of responses was significant both for psychology and economics students and it indicates that the most frequently selected answer in both cases (78.4% among psychology students and 68.4% among economics students) was the one that revealed an ideal representation of the data (mistake 2), that is, the representation of a trend or theoretical representation without the variability inherent to a real experiment and which in this case reveals a perfect linear relationship.

Finally, we wish to point out that there are very few responses that reveal serious misinterpretations (mistakes 3 and 4) and that these are mainly focused on problems 3 and 4 among both groups of students.

CONCLUSION

The picture of social science students in Spain revealed by the data we have just reported shows us that they are capable of relating textual information and certain graphical representations, but they also reveal significant shortcomings in their understanding about important syntactic aspects of this system of representation, which may have an effect on their professional skills or academic future. When interpreting our results, we should consider that students from our sample, especially economics students, had received abundant formal instruction on data analysis and representations. More specifically, our results indicate that a majority of these students do not relate the graphical representational formats to the nature of the variables represented (categorical, continuous) and that their processing is focused mainly on factual or explicit decoding of data (see, Curcio, 1987; Friel et al., 2001; Pérez-Echeverría, Martí, & Pozo, 2010; Pérez-Echeverría & Scheuer, 2009; Postigo & Pozo, 2000), in such a way that a kind of correspondence rule between each data point and their graphical representation is used. According to this rule, a graph is correct if there is a correspondence between each piece of information and each point in the graph, without taking other factors into account. From this point of view, it is easier to read (or to select as better graphical representation) a bar graph than a line graph (Wainer, 1980, 1992; Wavering, 1989). Therefore, there is no overall reading of the trends and relationships expressed graphically that enables relating these data to more theoretical concepts or situations.

This kind of correspondence rule is also reflected in problem 2, which reveals a different selection criterion from the other problems regarding the idea that a certain result should be a faithful reflection of or correspondence to the theory it tries to explain. In other words, in problem 2 the majority of students do not choose the original graph that had been proposed by researchers because it reveals some "sawtooth" waveforms that apparently do not reflect the overall relationship reported by the author, which describes a clean pattern without highlighting the exceptions to the pattern. We are currently interviewing psychology students about graphical selection in another study. Our preliminary results show that most students reject "sawtooth" graphs because they "do not correctly depict the results." According to this position, any apparent contradiction to a rule invalidates a theory. This situation reminds us of the representativeness heuristic reported by Tversky and Kahneman (1974).

These results contrast with those found by other researchers (see, for example, Gagatsis & Shiakalli, 2004), who, from tasks in which psychology students had to transform relatively simple textual information or algebraic expressions into graphs and vice versa, argue that the statistical training of these students enables them to make a functional and epistemic use of the graphs. The results also contrast with results we found in other tasks similar to those used by Gagatsis and Shiakalli (Perez-Echeverría et al., 2007; Pérez-Echeverría, Postigo, & Pecharromán, 2009) with psychology students with similar training to those who participated in this study and whose performance in the transformation of functions represented algebraically and through textual data (quasi-tables) into graphs and vice versa was suitable.

In our opinion, these contrasts are the result of the differences between reading/interpreting and constructing a graph. The previous research cited (e.g., Gagatsis & Shiakalli, 2004; Pérez-Echeverría et al., 2007) asked participants to translate organized data into a graph, but the task only required a translation from numerical data, point by point, into a representation. It was not necessary to interpret the data to complete the task. Consequently, achievement was very high because the task implied just a factual interpretation of data.

Our data, with the exception of problem 2, confirm the relationship between the difficulty of the problem and the characteristics of the variables (the kind of variable, number of variables, and complexity of these, and so on). However, our data do not appear to confirm the relationship between the degree of knowledge about the task's content and the way of perceiving or interpreting the graphs. There were no significant differences between the way in which psychology students interpreted the graphs that represented psychological information and those that showed another kind of content. Nor were there important differences between the performance of psychology and economics students. Our interpretation of these results is related both to the kind of tasks we have proposed and to the level of interpretation made by the students. In our task, we only asked for a selection of a graph and not the use of the graph, for example, in an argument or a prediction task in which we can expect that the content might have more of an influence. In fact, in a set of tasks in which we used the same graphs as those used in the study here reported (Pérez-Echeverría, Postigo, & Marín, in preparation) and in which students from the same fields and training were asked to interpret the results, we found some differences related to the psychological versus nonpsychological content compared with what we have just reported. However, it is also possible that the effect of the content on the interpretation of results is more obvious when reading the graph more in-depth, something not possible in the study we report here because we asked for a selection, not an interpretation of results.

Finally, we would like to stress that these results reveal that the performance of these social science students is probably a reflection of the methods used to teach about the use of graphical representation systems. These methods have an effect on decoding rules or construction of these systems but not on their conceptual meaning or functionality.

ACKNOWLEDGMENTS

This research was possible thanks to the Ministry of Education and Science in Spain grant (EDU2010-21995-C02-01) directed by J. I. Pozo. The authors also want to thank Bárbara M. Brizuela and Andee Rubin, reviewers of the chapter, for their comments.

REFERENCES

Balchin, W., & Coleman, A. (1965, November). Graphicacy should be the fourth ace in the pack. *The Times Educational Supplement*, p. 5.

Butler, D. L. (1993). Graphics in psychology: Pictures, data, and especially concepts. *Behaviour Research Methods, Instruments and Computers, 25*, 81–92.

Carpenter, P. A., & Shah, P. (1998). A model of perceptual and conceptual processes in graph comprehension. *Journal of Experimental Psychology: Applied, 4*, 75–100.

Carswell, C. M., Emery, C., & Lonon, A. M. (1993). Stimulus complexity and information integration in the spontaneous interpretation of line graphs. *Applied Cognitive Psychology, 7*, 341–357.

Curcio, F. R. (1987). Comprehension of mathematical relationships expressed in graphs. *Journal for Research in Mathematics Education, 18*, 382–393.

Donald, M. (1991). *Origins of the modern mind. Three stages in the evolution of culture and cognition.* Cambridge, MA: Harvard University Press.

Eisner, E. W. (1974). Is the artists-in-the-schools program effective? *Art Education, 27*(2), 19–23.

Eisner, E. W. (1994). *Cognition and curriculum reconsidered.* New York: Teachers College Press.

Friel, S. N., Curcio, F. R., & Bright, G. W. (2001). Making sense of graphs: Critical factors influencing comprehension and instructional implications. *Journal for Research in Mathematics Education, 32*(2), 124–158.

Gabucio, F., Martí, E., Enfedaque, J., Gilabert, S., & Konstantinidou, K. (2010). Niveles de comprensión de las tablas en alumnos de primaria y secundaria. *Cultura y Educación, 22*(2), 183–197.

Gagatsis, A., & Shiakalli, M. (2004). Translation ability and problem solving. *Educational Psychology, 24*(5), 645–657.

Gelman, A., Pasarica, C., & Dodhia, R. (2002). Let's practice what we preach: Turning tables into graphs. *The American Statistician, 56,* 121–130.

Gimeno Sacristán, J. (2007). *El curriculum: Una reflexión sobre la práctica.* Madrid, Spain: Morata.

Kosslyn, S. M. (2006). *Graph design for the eye and mind.* New York: Oxford University Press.

Latour, B. (1990). Drawing things together. In M. Lynch, & S. Woolgar (Eds.), *Representation in scientific practice* (pp. 19–68). Cambridge, MA: MIT Press.

Lévi-Strauss, C. (1963). *Structural anthropology.* New York: Basic Books.

Martí, E. (2003). *Representar el mundo externamente. La adquisición infantil de los sistemas externos de representación.* Madrid, Spain: A. Machado Libros.

Martí, E., Gabucio, F., Enfedaque, J., & Gilabert, S. (2010). Cuando los alumnos interpretan un gráfico de frecuencias. Niveles de comprensión y obstáculos cognitivos. *Revista IRICE, 21,* 65–81.

Martí, E., & Pozo, J. I. (2001). Más allá de las representaciones mentales: La adquisición de los sistemas externos de representación. *Infancia y Aprendizaje, 90,* 11–30.

Meyer, J., Shinar, D., & Leiser, D. (1997). Multiple factors that determine performance with tables and graphs. *Human Factors, 39*(2), 268–286.

Olson, D. (1994). *The world on paper.* Cambridge, UK: Cambridge University Press.

Peden, B. F., & Hausmann, S. E. (2000). Data graphs in introductory and upper level psychology textbooks: A content analysis. *Teaching of Psychology, 27*(2), 93–97.

Pérez-Echeverría, M. P., Martí, E., & Pozo, J. I. (2010). Los sistemas externos de representación como herramientas de la mente. *Cultura y Educación, 22*(2), 133–147.

Pérez-Echeverría, M. P., Pecharromán, A., & Postigo, Y. (2007). Los sistemas de representación externa en el aprendizaje: La habilidad para traducir información a distintos formatos. In J. I. Pozo, & F. Flores (Eds.), *Cambio conceptual y representacional en el aprendizaje y la enseñanza de la ciencia* (pp. 107–122). Madrid, Spain: A. Machado Libros.

Pérez-Echeverría, M. P., Postigo, Y., López Manjón, A., & Marín, C. (2009). Aprender con imágenes e información gráfica. In J. I. Pozo & M. P. Pérez-Echeverría (Eds.), *Psicología del aprendizaje universitario: La formación en competencias* (pp. 134–148). Madrid, Spain: Morata.

Pérez-Echeverría, M. P., Postigo, Y., & Marín, C. (2010). Las habilidades gráficas de los estudiantes universitarios: ¿Cómo comprenden las gráficas los estudiantes de psicología? *Cultura y Educación, 22*(2), 215–229.

Pérez-Echeverría, M. P., Postigo, Y., & Marín, C. (In preparation). *Comprender e interpretar gráficas para su uso en el aprendizaje universitario.*

Pérez-Echeverría, M. P., Postigo, Y., & Pecharromán, A. (2009). Graphicacy: University students' skills in translating information. In C. Andersen, N. Scheuer, M. P. Pérez-Echeverría, & E. V. Teubal (Eds.), *Representational systems and practices as learning tools* (pp. 209–224). Rotterdam, Netherlands: Sense Publishers.

Pérez-Echeverría, M. P., & Scheuer, N. (2009). External representations as learning tools. In C. Andersen, N. Scheuer, M. P. Pérez-Echeverría, & E. V. Teubal (Eds.), *Representational systems and practices as learning tools* (pp. 209–224). Rotterdam, Netherlands: Sense Publishers.

Postigo, Y., Pérez-Echeverría, M. P., & Marín, C. (2010). ¿Cómo usan y comprenden los gráficos los estudiantes universitarios? *Revista IRICE, 21,* 81–94.

Postigo, Y., & Pozo, J. I. (2000). Cuando una gráfica vale más que 1000 datos: Interpretación de gráficas por alumnos adolescentes. *Infancia y Aprendizaje, 90,* 89–110.

Romberg, T. A., Fennema, E., & Carpenter, T. P. (Eds.). (1993). *Integrating research on the graphical representation of functions.* Mahwah, NJ: Lawrence Erlbaum Associates.

Roth, W. M., & Bowen, G. M. (2003). When are graphs worth ten thousand words? *Cognition and Instruction, 21*(4), 429–473.

Roth, W. M., Pozzer-Ardenghi, L., & Han, J. (2005). *Critical graphicacy: Understanding visual representation practices in school science.* Dordrecht, Netherlands: Springer.

Shah, P., Freedman, E. G., & Vekiri, I. (2005). The comprehension of quantitative information in graphical display. In P. Shah & A. Miyake (Eds.), *The Cambridge handbook of visuospatial thinking* (pp. 426–477). Cambridge, UK: Cambridge University Press.

Stern, E., Aprea, C., & Ebner, H. G. (2003). Improving cross-content transfer in text processing by means of active graphical representation. *Learning and Instruction, 13*(2), 191–203.

Tversky, A., & Kahneman, D. (1974). Judgment under uncertainty: Heuristic and biases. *Science, 185,* 1124–1131.

Wainer, H. (1980). Making newspaper graphs fit to print. In P. A. Kolers, M. E. Wrolstad, & H. Bouma (Eds.), *Processing of visible language* (vol. 2, pp. 125–142). New York: Plenum.

Wainer, H. (1992). Understanding graphs and tables. *Educational Researcher, 21*(1), 14–23.

Wavering, M. J. (1989). Logical reasoning necessary to make line graphs. *Journal of Research in Science Teaching, 26,* 373–379.

CHAPTER 5

Eppur Si Muove

The Representation of the Intrinsic Movement of Matter

Miguel Angel Gómez Crespo,

Juan Ignacio Pozo, and

María Sagrario Gutiérrez Julián

As some recent studies have shown, understanding of the kinetic molecular model, one of the essential contents in secondary school science, is a difficult task for students and requires a complex process of conceptual change. According to these studies, which confirm trends previously found (e.g., Pozo, Gómez Crespo, & Sanz, 1999; Stavy, 1995), students come to accept the academic proposition that matter is composed of particles that they cannot see, but when they are confronted with everyday situations, they fall back on their intuitive ideas, which are more consistent with properties that can be observed in the macroscopic world that surrounds them. These studies seem to indicate that, although the idea that matter is made up of tiny particles is readily accepted, students do not seem to recognize it as something inherently powerful and useful for understanding and talking about everyday events. Indeed, students do not readily refer to particles, intrinsic motion, or vacuum in their explanations (Meheut, 2004). The probability of use of the molecular model, then, depends on other variables, such as the format of the task, the context within which it is framed, its complexity, or the type of phenomenon being studied (Pozo, Gómez Crespo, & Sanz, 1999; Williamson, Huffman, & Peck, 2004). Taking this into account, the fact that students refer to particles more frequently does not necessarily mean that they have a better understanding of the molecular model.

What does vary with instruction, and with the consequent conceptual change, if change in fact takes place, is the interpretation students make of particles within the kinetic molecular model when they need to use the model. In general, the notions of conservation, applied at a microscopic level (e.g., conservation of size, number of particles, when water changes to ice) pose fewer problems for secondary school students than the comprehension of the mechanisms and relations between particles (e.g., explaining how water becomes ice; see Gómez Crespo & Pozo, 2004; Meheut, 2004). Students tend to assign erroneous properties to particles, using ideas that are, in many cases, closer to their own intuitions than to scientific principles. In general, students attribute the same properties to particles as they do to the system that they are a part of (e.g., when the pressure inside a balloon increases as it expands, the particles that compose air accumulate at the walls). Thus, as students increase their use of the kinetic molecular model, instead of re-describing their intuitive ideas in scientific terms, they tend to reinterpret scientific concepts in terms of the same intuitions that constrain their understanding of these chemical concepts (Pozo & Gómez Crespo, 2005).

According to our theoretical framework, all these conceptual difficulties in learning elementary chemistry are in part due to the persistence and pervasiveness of students' intuitive representations of matter, their *intuitive chemistry*. We assume that as a result of our biological and cultural constraints, human beings spontaneously construct sets of intuitive representations (i.e., mental models or simulations) about both psychological and physical objects. These representations are essentially implicit, as generally people are not conscious of them and thus cannot communicate or explain them; but, above all, they are implicit because they are based on assumptions or implicit principles that constrain the construction of these intuitive representations and give them the form of intuitive or implicit theories (Chi, 2008; Vosniadou, Vamvakoussi, & Skopeliti, 2008). These representations share some cognitive traits across different domains, such as psychology, as in students' ideas about the learning process (e.g., Bautista, Pérez Echeverría, Pozo, & Brizuela, 2012), or physics, as in ideas about the nature of vision and color (Bravo, Pesa, & Pozo, 2009) or about the nature of matter (Gómez Crespo & Pozo, 2004). For the purposes of this chapter, one of these common constraints is an epistemological realism that implicitly assumes a similarity between these representations and the perceived or sensed world (Vosniadou, Vamvakoussi, & Skopeliti, 2008). As a consequence, our intuitive representations have an embodied nature, as they reflect how our body constrains, but also empowers, our representation of physical and psychological objects. Within this embodied approach to representations (e.g., Gibbs, 2006; Glenberg, 1997), according to Damasio (1994), our representations do not reflect how the world changes but how the world changes our bodily actions and perceptions. In the case of chemistry, this embodied nature would reflect the way we perceive and interact with matter in its different states (solid, liquid, and gaseous) starting with our first months of life.

In fact, according to Spelke (1994), 3-month-old babies already have some principles that they use to represent the physical world in which they live, in such a way that the baby's physical world is made of solid objects that only move if another object or agent acts upon it and whose parts move in unison following defined trajectories. But the embodied representation of liquids, and especially gases, should be different as its surface appearance is different from that of solids. Liquids, and especially gases, can be easily disaggregated, dissolved, and so forth, and they frequently seem to be in macroscopic motion without an apparent external cause. Thus, students could have an embodied and macroscopic representation of matter as "solid stuff" and they may not differentiate between this level and a microscopic analysis of matter. They are apparently attributing intrinsic movement to gases but not to solids. According to this interpretation, it is interesting to wonder, as we will show later, how students would represent liquids according to its apparent movement (e.g., the bubbles in a soft drink or still water in a glass).

A second constraint on the intuitive representation of matter is based on the conceptual structure implicitly assumed. In general, our intuitive ideas in different domains assume simple agent–object structures (i.e., when a cause leads to a direct and linear effect in an object, such as when an object moves faster due to a related increase in the supposed impulse force). However, in order to interpret the relations between particles, the kinetic molecular theory relies on the scheme of interaction, as is true in other scientific domains (Chi, 2008). Interaction requires considering the influence of certain variables in more complex conceptual structures than linear agent–object relations, as these different factors or variables influence one another, in a cyclical or interactive system.

All these constraints result in students having difficulty interpreting the relations among particles in terms of interactions within a system. As a consequence, they have difficulties understanding three fundamental notions, which go against alternative theories that are based on the observable appearance of matter: the discontinuity of matter (along with the notion of vacuum), the intrinsic motion of the particles that make up matter, and the mechanisms implied in the changes of the state of matter.

Within this framework, this chapter will be centered on one of these essential notions for secondary school science: the understanding of the intrinsic movement of the particles that make up matter, an idea very difficult for students to grasp (e.g., Abdo & Taber, 2009). Thus, some students have strong and persistent intuitive representations, based on perceptions of our everyday *macroscopic* world that are counter to the idea that particles are in continuous movement and interaction. Assuming the embodied nature of these representations, matter would be perceived as inert and at rest, unless some agent interacts with it. These intuitive embodied representations persist throughout adolescence and even among university students with specific knowledge of chemistry (Pozo & Gómez Crespo, 2005). Learning chemistry, as is true with conceptual change in other domains, does not involve a replacement of intuitive representations with scientific knowledge (Pozo, Gómez Crespo, & Sanz, 1999). Instead, it involves the use of a plurality of representations for the same object or domain, a coexistence, not always conscious, of both kinds of representations, the implicit/embodied intuitions and the explicit/formalized knowledge. Naïve theories about movement are quite persistent and instruction does not seem to easily modify these theories (Pozo & Gómez Crespo, 2005), failing frequently to make the scientific model of matter (based on continuous intrinsic movement) understandable or believable. This, in turn, makes new instructional approaches necessary in order to hierarchically integrate or redescribe the intuitive representations in scientific terms, as we will argue in the last section of this chapter.

HOW DO STUDENTS CONCEIVE THE MOVEMENT OF PARTICLES?

Within this theoretical framework, the resistance to accept the intrinsic motion of particles should vary according to the aggregate state of matter (e.g., Gómez Crespo & Pozo, 2004; Stavy, 1995). In the case of movement, it seems easier for students to attribute the idea of intrinsic motion to gases and liquids than it is for solids (Abdo & Taber, 2009). Students' intuitive representations of the different aggregate states could be accounted for by the fact that they do not differentiate between the intrinsic movement of the particles that compose a material and the apparent movement of the same material (i.e., its perceived movement).

Following our previous studies on the learning of the nature of matter, this chapter analyzes the most frequent representations of the movement of particles that compose matter, which is the pattern of activation for these representations according to the educational level of the students and the specific science instruction they have received. We are also interested in the influence of the macroscopic appearance of matter (i.e., solid, liquid, or gaseous) on students' representational patterns and will therefore explore the main paths of progression in the learning of the intrinsic motion of particles as an essential component of chemistry learning in secondary and postsecondary science teaching. More specifically, we seek to do the following:

1. To compare the representations of movement in the constituent particles of matter among students of different ages and levels of education in chemistry.
2. To analyze how the macroscopic or apparent movement of a studied substance influences the activation of the different representations of the nature of matter the students may have.
3. To analyze the most common representations that students have regarding movement, and to look at how they are affected by the level of education and the aggregate state of a given substance (solid, liquid, or gas).

We designed a questionnaire that included 12 multiple-choice items, each of which posed four tasks, each item corresponding to one of the three states of matter (solid, liquid, and gas) that were studied in our previous research (Pozo & Gómez Crespo, 2005). In all questions, a phenomenon that was assumed to be familiar to or easily understood by the students was presented. For each item, we always presented the same response categories (see Figure 5.1), which corresponded to four different representations of the behavior of matter. Thus, Option A implied a totally static representation (reflecting the everyday perception of matter at a macroscopic level; according to studies on intuitive physics, we assumed that some students would understand objects as being "inanimate"); Option B implied a static representation with extrinsic movement, caused by an external agent (according to these intuitions, we assumed that some students would hold that objects only move if another moving object comes into contact with them); Option C implied an intrinsic and continuous movement that chemistry attributes to particles (the model assumed by the kinetic molecular theory); and Option D implied an extrinsic movement, but caused by an internal agent (when we attribute the movement to a macroscopic action perceived within the system or the object). The 12 items were distributed in the questionnaire following a random criterion.

We first analyzed the relationship between the different variables manipulated and the frequency of correct responses. Second, we analyzed how the participants used the different response categories and the proportion in which each response category was used for each one of the variables studied.

Figure 5.1. Example of Questionnaire Items

The windows of your classroom, like all windows, have a pane of glass. In what state do you think the particles are that make up the glass in the window?

    A. They are always still, stationary.
    B. They only move if we shake the glass.
    C. They are always moving.
    D. They only move if there are air bubbles that push them.

A little girl is playing with a balloon. The man who sold her the balloon filled it with a gas called helium and then closed the opening with a knot. In what state do you think the particles that make up the helium gas inside the balloon will be?

    A. They are always still, stationary.
    B. They only move if we shake the balloon.
    C. They are always moving.
    D. They only move when the air inside the balloon pushes them.

We administered this questionnaire to 219 Spanish students of different ages and educational levels distributed in five groups. Three of the groups were made up of adolescents of different ages, with different levels of instruction in chemistry, and drawn from the same secondary school. The younger group, which had received less instruction in chemistry, was composed of 60 9th-graders (14 to 15 years old); they had studied the kinetic theory within their science curriculum, based on three main ideas—namely, that matter is composed of particles, that these particles are in continuous movement, and that they are separated by a vacuum. The two other adolescent groups were composed of 11th-graders (16 to 17 years old). They differed only in the instruction they had received in science. The first group of 11th-graders (16–17 Science [S]) consisted of 36 participants who selected science studies and thus received 2 additional years of instruction in physics and chemistry, while the 53 students without science studies (16–17 non Science [non S]) did not receive this additional instruction in science. The two older groups were made up of university students also with and without studies in science. All of them had already graduated from 5-year universities in Madrid and were students in a pre-service teacher course, which in Spain was a requirement for obtaining a position as secondary school teacher in a specific subject-matter domain. Our sample in these groups was composed by 17 graduates in chemistry or physics studies (Univ-S) and 53 graduates in social science or philosophy studies (Univ-non S). Although the chemistry or physics graduates cannot truly be considered experts in chemistry, in this study they could be considered experts, as they should have sufficient knowledge to solve the elementary chemistry tasks used in this study. In fact, in a few months many of them would possibly be teaching these same concepts we were interviewing them about to secondary students. In contrast, the nonscience graduate group was considered the novice group because they had no specific training in chemistry. In fact, their level of instruction in science, specifically in physics and chemistry, was similar to that received by the group of 16- to 17-year-old students without science studies (16–17 non S).

DO STUDENTS REALLY UNDERSTAND THE MOVEMENT OF PARTICLES?

First of all, we can see an effect of age and level of instruction on the performance obtained on the questionnaire, showing a significant effect for the variable *group* [$F(4,214) = 5.478$; $p < 0.001$]. As reflected in the proportion of correct responses for each group shown in Table 5.1, there were significant differences between the two most extreme groups. Younger adolescents (14–15) used the kinetic molecular model less than older adolescents with specific science instruction and university students with science studies, but also less than students without specific instruction in science. Moreover, the Univ-Science group scored above the 16–17 non S group. Thus, a relevant result of this study is that science teaching alone does not seem to lead to differences in students' understanding of the intrinsic movement of particles.

The analysis also shows that students' representations were different according to the apparent *state of matter* [$F(2,428) = 107.701$; $p < 0.001$] (see Table 5.2), varying indeed for the three states of matter ($p < 0.001$). The intrinsic motion was attributed more easily to gases (0.87) than to either liquids (0.69) or solids (0.48). This is consistent with the hypothesis that the theory used to explain the motion of a material depends on its perceptual appearance. It is easier to assume intrinsic motion in gases, which appear more "ethereal," than in solids, which have a more inert appearance. For liquids, we found an intermediate pattern.

This effect is clearer if we take into account the analysis of the interaction between the two main variables, group and state of matter [$F(8,428) = 2.898$; $p = 0.004$]. The proportion of correct responses for each group and state of matter can be seen in Table 5.2. The differences between groups are significant ($p < 0.001$) and follow the trends that were previously mentioned (i.e., that intrinsic motion can be more easily accepted in the case of gases). The same pattern can be seen with the general scores, in keeping with the hypothesis regarding the influence of perceptual appearance. There is, however, an exception to this trend in the group of experts (Univ-S). Although we can observe in this group the same general trend, there were no differences in the attribution of intrinsic movement to the particles of gases, liquids, and solids. This result confirms that the use of scientific models provides a higher representational consistency across tasks and contexts (Pozo & Gómez Crespo, 2005).

We also considered interaction between group and state of matter. In the case of solids, results were very similar to the general pattern identified above, as the only differences found were between the most extreme groups. Once again, secondary science does not seem to change students' intuitive representations: They

Table 5.1. Proportion of Correct Responses for Each Group

	14–15	16–17 non S	16–17 S	University students (non S)	University students (S)
Proportion	0.53	0.61	0.70	0.71	0.86

Table 5.2. Proportion of Correct Responses for the Interaction Group x Content

	14–15	16–17 non S	16–17 S	University students (non S)	University students (S)
Solids	0.29	0.35	0.47	0.55	0.77
Liquids	0.50	0.64	0.70	0.72	0.87
Gases	0.81	0.83	0.93	0.85	0.94

assume that solid objects, like static entities, only move as a result of an external agent acting upon them. When considering liquids, we can see smaller differences among groups than with solids. The only differences were between younger adolescents (14–15) and the two university groups. In gases, where the use of the kinetic model was more frequent, there were no differences among groups, as all of them apparently accepted its intrinsic movement.

Once again, it is worth highlighting that within the same educational level there were no differences across the three states of matter (i.e., solid, liquid, and gas) as a consequence of the specific science instruction received, both between adolescent groups (16–17 non S and 16–17 S) and between university students (Univ-S and Univ-non S). Although specific science teaching seemed to lead to an increase of the intrinsic movement answers, the effect was not statistically significant. This scarce effect of science teaching is especially noteworthy for answers regarding the solid state, where the 16–17 S group's use of the kinetic model was clearly poor. Even among Univ-S we found that the static representation of solids was used more frequently than what would be expected. This should be surprising, as these students have a high level of education in chemistry. However, before we go into this problem further, we will first look at the analysis of the response categories used by the students.

WHAT REPRESENTATIONS WERE MORE FREQUENT?

The four response categories were designed to differentiate not only between the scientific model (intrinsic movement) and intuitive representations, but also within these intuitive representations among different models (at rest, external cause, and internal cause) according to the state in which matter is presented (solid, liquid, and gas). We used an ANOVA (analysis of variant) to look at the interaction *group x state of matter x category*. We will only describe the results obtained for the variable *category* and the interaction among all the variables. The results obtained with the other main variables (*group* and *state of matter*) are similar to the results seen previously in this chapter.

Students use some kinds of representations more often than others [$F(3,856)$ = 288.584; $p < 0.001$]. The most frequently used category (65% of the cases) is the one that we considered scientific (the intrinsic movement of particles). The category "at rest," the absence of movement in the particles, was used in 16% of the cases, and was used more than the rest of the categories. Extrinsic movement caused by an external agent (11%) and extrinsic movement caused by an internal agent (9%) were used in a smaller proportion.

Nevertheless, keeping in mind the interaction between *groups* and *states of matter*, differences appear that will help clarify the form in which they are used. A significant effect was found for the interaction *group x category* [$F(12,856)$ = 5.021;

$p < 0.001$]. As can be seen in Table 5.3, the most frequent response continues to be the acceptance of the intrinsic movement of particles.

The most frequent intuitive representation was the *at rest* category. In fact, there were no differences in the use of this commonsense representation as a consequence of age and instruction. All the groups used it in a similar proportion, showing that its relatively moderate presence is nevertheless very persistent and resistant to instruction. Concerning the attribution of movement to the action of an *external agent*, we found that younger adolescents (14–15) used this representation more frequently ($p < 0.01$) than the remaining groups, without differences between them. Finally, the reference to an *internal agent* as the cause of the movement of particles is higher ($p < 0.001$) among 14–15 and 16–17-non S adolescents, the younger groups and with less instruction in science, while the other adolescent group, with more instruction in science (16–17-S), hardly refers to this idea.

But are these alternative representations of movement distributed equally among all the states of matter? Studying the interaction of all three factors may shed some light on the situation. In fact, this interaction *group x state of matter x category* was found to be significant [$F(24,856) = 2.319; p < 0.001$], and thus it can help clarify how each group uses the different response categories as a function of the state of matter.

As can be seen in Table 5.4, for gases, the most frequent interpretation in all groups, without differences among them, was to attribute an intrinsic movement, that is to say, the correct response, as found in previous studies. This response was chosen for the majority of answers from the group 14–15 up to and including the group of experts. The alternative categories, as can be imagined, were used in very low proportions when considering gases, with some differences among groups.

When considering *liquids*, the most frequently used category in all the groups is again intrinsic movement (see Table 5.5), the differences between groups being really similar to the global pattern just described in Table 5.3, and showing thus a scarce effect of age and instruction on the understanding of liquids. There were, however, a certain number of responses that attributed the movement of liquids

Table 5.3. Average Proportion, in Each Group, of the Use of Every Response Category

	14–15	16–17 non S	16–17 S	University students (non S)	University students (S)
At rest	0.18	0.15	0.14	0.15	0.11
External agent	0.16	0.09	0.10	0.10	0.03
Intrinsic movement	0.53	0.61	0.70	0.71	0.86
Internal agent	0.13	0.15	0.06	0.04	0

Table 5.4. Average Scores Obtained for Each Group
and Category for the Content Gases

	14–15	16–17 non S	16–17 S	University students (non S)	University students (S)
At rest	0.02	0.02	0.01	0.06	0.06
External agent	0.08	0.05	0.03	0.09	0
Intrinsic movement	0.81	0.83	0.93	0.85	0.94
Internal agent	0.09	0.10	0.03	0.01	0

Table 5.5. Average Scores Obtained for Each Group
and Category for Liquids

	14–15	16–17 non S	16–17 S	University students (non S)	University students (S)
At rest	0.03	0.02	0	0.07	0.06
External agent	0.29	0.16	0.23	0.16	0.07
Intrinsic movement	0.50	0.64	0.70	0.72	0.87
Internal agent	0.18	0.18	0.07	0.06	0

to external or internal agents. These categories did not attribute movement to the intrinsic nature of matter, but instead to the causal action of some agent (e.g., the liquid only moves when the container is shaken). It is noteworthy, however, that the groups that most frequently explained the liquid state in terms of external agents were the two university groups. The only differences for this category are, however, due to its less frequent use in the 14–15 group ($p < 0.01$).

Concerning the attribution of movement to an *intrinsic agent*, we observe (see Table 5.5) that the younger groups that had received less science instruction (14–15 and 16–17 non S) were more prone to this explanation. On the other hand, 16–17 S and both groups of university students hardly used it. Only the expert group was able to avoid this intuitive theory completely. This result led us to analyze the possible influence of the external appearance of the liquid. For this purpose, we compared the items that make a reference to liquids at rest and the items that involved liquids with an apparent external motion. Qualitatively, in the items that refer to liquids that have the appearance of motion (Coca-Cola and effervescent aspirin), students were more likely to attribute the motion to an internal agent (see

Figure 5.2). On the other hand, when the liquids were apparently at rest (oil and water in a glass), the reference to an external agent to explain the movement of the particles was more frequent. This effect was observed less frequently among the younger groups with less science instruction. Once again, the intuitive representations about the movement of particles seem to be of an embodied nature (Pozo & Gómez Crespo, 2005), as they do not differentiate between the apparent and the intrinsic motion of matter, reflecting once again the constraints that our body imposes on our perception of physical objects. According to Glenberg (1997), Gibbs (2006), and Damasio (1994), a representation is embodied when it is constrained by the way our body interacts with that object. In this approach, you do not need to study body movements in order to analyze embodied representations, because, in fact, the body is in the mind (or the mind has a bodily structure).

In summary, the representations that are most common and persistent, with respect to liquids, consist of attributing the possible movement to an immediate and causal action. There are hardly any responses that refer to particles that are immobile or at rest. Instead, it is assumed that the particles move only when they are acted upon by an external or an internal agent.

Concerning *solids*, as shown in Table 5.6, the two most frequent interpretations are absolute rest and intrinsic or continuous motion, although in some groups the proportions of answers in the other categories also reach levels that make their consideration relevant.

Concerning the *at rest* representation, there were no significant differences between any adolescent group, as all of them appealed very often to this intuitive

Figure 5.2. Mean Attribution for Each Group of Particle Movement in Liquid to External and Internal Causes, Differentiating the Liquids with and without an Apparent Movement

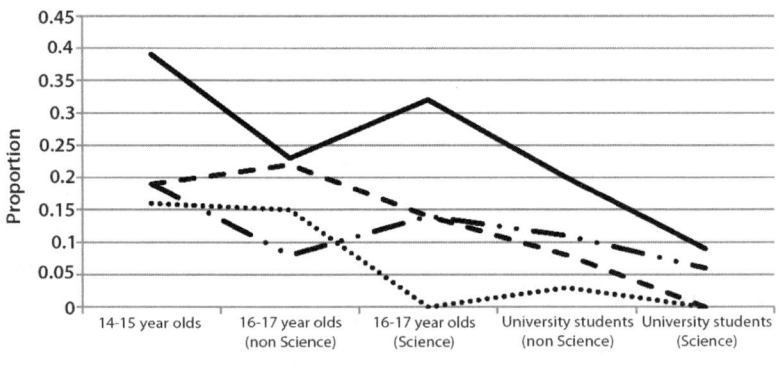

Table 5.6. Average Scores for Group and Content with Solids

	14–15	16–17 non S	16–17 S	University students (non S)	University students (S)
At rest	0.48	0.42	0.41	0.35	0.22
External agent	0.10	0.06	0.03	0.05	0.02
Intrinsic movement	0.29	0.35	0.47	0.55	0.77
Internal agent	0.13	0.18	0.09	0.05	0

representation. The university groups were less inclined, in some cases significantly less, to accept the idea that particles in solids are static, but it was still the second most frequently used category for these groups after the scientific model. In fact, this idea seems to be the truly *alternative* representation to the scientific model for all the groups. Science learning decreases its use, but once again it does not lead to its disappearance.

When referring to alternative explanations of motion based on the action of an internal or external agent, we can see that their use is less frequent (see Table 5.6). Despite this, the explanation in terms of an internal agent appears in a lower proportion, with the exception of the experts, who never use it. The interpretation that considers movement as a consequence of the action of an external agent is the least used of all four of the answer categories.

To sum up, for solids, we find that there is a highly accentuated tendency to interpret the state of their constituent particles as being at absolute rest. Although this tendency decreases with age, it does not disappear. Once again, specific science instruction does not significantly reduce this idea, as there are no differences in its use between adolescents with and without science studies (16–17 S and 16–17 non S groups). This is still a persistent idea even among university students with training in science; it is more frequent than should be expected and, above all, desired, as these students could be, in a few months, science teachers for adolescent students similar to those included in this study.

CONCLUSIONS ABOUT STUDENTS' UNDERSTANDING OF PARTICLE MOVEMENT

Returning to the three main objectives of our investigation into students' understanding of the intrinsic movement of particles, regarding *the influence of age and instruction* on students' knowledge, the data show a more frequent use of the scientific theory, based on the continuous and intrinsic movement of particles, as age and education increase. Nevertheless, within this general trend, the effect of specific instruction in science is less than can be expected or desired. Although the Univ-S group has a higher performance than the remaining groups, they continue using nonscientific representations quite often, especially when matter is in the solid state. This fact reinforces the hypothesis that learning science does not imply abandonment but rather a multiplicity of representations (Pozo, Gómez Crespo, & Sanz, 1999). In fact, in all the groups studied, there were many alternative representations, which were not always differentiated.

In addition, and even more surprising, is the discouraging fact that adolescents who have received more instruction in science (16–17 S) show a performance that is systematically very close to that of the students of the same age who have not studied science (16–17 non S). Even the science university students, who may end up being science teachers, have a representation of particles too close to that of university students without specific education in science. It appears that studying science does not lead to a generalized conceptual change in the understanding of matter, as it does not help to differentiate between intrinsic motion within particles and apparent motion, a result we found in a previous study (Pozo & Gómez Crespo, 2005).

Concerning our second objective, the *influence of the state of matter* is a determining variable in the representation activated among the subjects of all groups. In all groups, regardless of age and instruction in chemistry, it is easier to understand the intrinsic motion of particles when they are in a gaseous state than when they are liquids. Likewise, it is easier to attribute intrinsic motion to the particles in liquids than in solids. This conclusion, in the case of adolescents, confirms the data obtained by previous studies (e.g., Stavy, 1995).

The study of the *use of different alternative representations*, the third objective of this study, shows that across all conditions, the category most used is one that corresponds to the acceptance of intrinsic motion. Nevertheless, in approximately one-third of the cases, subjects continue to use intuitive representations, depending on the appearance of matter (i.e., its physical state). We find that it is easier to accept the movement of particles in gases, which are found in nature in a more changing, dynamic, and "ethereal" form, than in other states of matter. In fact, it was the need to find an explanation for the behavior of gas that made the kinetic model of matter itself historically necessary. In contrast to this, when matter is presented in solid form, the idea that its particles do not move is more frequent. This is consistent with our own macroscopic perception of these substances. Liquids present an intermediate situation in which the acceptance of intrinsic movement is situated between the two extremes. For liquids, the intuitive representations depend on the state of apparent motion that is found. If the liquid is at rest, the possible movement of its particles is attributed to an external agent (e.g., someone who moved the container). However, if the liquid can be observed to be in motion, the movement of its particles is explained by some element that is in the liquid (e.g., the bubbles in the soft drink push the particles).

From this trend in the students' responses, we can conclude that the intuitive representations about movement have their origin in the macroscopic perception of matter in such a way that the students tend to attribute movement to the particles only when they perceive that a material has moved. This suggests, therefore, that students do not differentiate between macroscopic and microscopic but rather apply the associative rule of similarity such that the model (microscopic) and the reality that it represents (macroscopic) are supposed to have the same attributes. This is consistent with the hypothesis of the embodied origin of implicit representations (Pozo & Gómez Crespo, 2005).

The differences in interpretation between the different aggregate states or appearance of matter in the intuitive representations, in contrast to the *expletive coherence* of the scientific theory that explains all of these situations with a single model, are related to the conceptual structures that underlie many other types of explanation. *Intrinsic motion* is the product of a system of interactions and changes between particles that modifies the velocity of this movement according to the state of aggregation or the degree of interaction between these particles establishing a cycle of interactions. On the other hand, *apparent movement* is explained by a simplified causal linear agent–object model (Andersson, 1986), in which the external or internal agent changes the natural state of the object (absolute rest). This schema is similar to what the students use to interpret the macroscopic movement of objects in which the movement is the product of a causal action but the state of rest is conceived of as the natural state. According to this interpretation, the students will tend to attribute the movement of the particles to the action of other agents (e.g., a person who disturbs the glass, another substance that pushes it, and so on).

However, not all the difficulties in the comprehension of intrinsic motion can be accounted for by the origin and nature of alternative conceptions. It is possible that some kind of instruction in science could have a negative effect on students' representations, reinforcing their original intuitive representations, instead of helping them to achieve a differentiation between the macroscopic perception and the microscopic analysis of matter. As Abdo and Taber (2009) have also found, when students begin to approach the study of the properties of matter, they do so by way of the idea of chemical bonds and interpret the properties of substances based on them. Thus, when they study the state of solids (fundamentally that of crystals), they are taught that the relative position of the particles (ions, in this case) is maintained and the possibility that these particles vibrate around their positions of equilibrium is almost an afterthought. Something similar happens with chemistry students: The possibility that particles vibrate is quite real, but completely out of context in terms of the properties of solids except in very specific contexts (e.g., when they have to explain the mechanisms of changes of state; Gómez Crespo, & Pozo, 2004).

A specific analysis of the influence of diverse types of instruction over the comprehension of the kinetic model is needed. To promote conceptual change in science, we will need to implement instructional strategies to help students both make explicit their intuitive/implicit representations and to differentiate them from the scientific knowledge they are supposedly learning. However, this conceptual change does not require leaving out these intuitive representations but rather re-describing (Karmiloff-Smith, 1992) or hierarchically integrating them in more complex, formalized, structures of scientific knowledge (Bautista et al., 2012; Bravo, Pesa, & Pozo, 2009).

These results also confront us with new challenges and questions. How does this evolution from those alternative conceptions—based on a causal agent, which makes the particle move—to the scientific model that assumes intrinsic motion takes place? How do different teaching strategies aimed at promoting conceptual change in chemistry affect this evolution? One of the approaches we are now developing is the study of the different conceptual profiles found among students that could be useful in explaining different paths of change in the learning of these core concepts about the nature of matter. We are also looking in this way to contribute to the design of teaching strategies that could help us promote conceptual change—in terms of a representational redescription (Karmiloff-Smith, 1992)—in the learning of chemistry. Our goal is for students not simply to replace their intuitions with scientific models, but to be able to interpret or redescribe their everyday experience with solid, liquid, and gaseous matter in terms of the interaction of particles in continuous movement through a vacuum space. In other studies (e.g., Gómez Crespo, 2008), we have found that instructional strategies based on the differentiation and comparison of models (both macroscopic and microscopic) promote conceptual change in this domain. Moreover, an adequate use of external representation systems through computer simulations will make easier both the differentiation and the hierarchical integration of everyday and scientific models. The objective should be that students will be able to translate or redescribe their embodied and intuitive representations in terms of scientific concepts instead of, as usually happens, reinterpreting the scientific models in intuitive terms.

ACKNOWLEDGMENTS

This research was supported by the Spanish Ministry of Science and Innovation through the project EDU2010-21995-C02-01.

REFERENCES

Abdo, K., & Taber, K. S. (2009). Learners' mental models of the particle nature of matter: A study of 16-year-old Swedish science students. *International Journal of Science Education, 31*(6), 757–786.

Andersson, B. (1986). Pupils' explanations of some aspects of chemical reactions. *Science Education, 70*(5), 549–563.

Bautista, A., Pérez Echeverría, M. P., Pozo, J. I., & Brizuela, B. M. (2012). Music students' conceptions of learning and instruction: Do these conceptions form theoretically consistent profiles? *Estudios de Psicología, 33*(1), 79–104.

Bravo, B., Pesa, M., & Pozo, J. I. (2009). The learning of sciences: A gradual change in the way of learning. The case of vision. *Investigaçoes em ensino de ciencias, 14*(2), 299–317.

Chi, M. T. H. (2008). Three types of conceptual change: Belief revision, mental model transformation, and category shift. In S. Vosniadou (Ed.), *International handbook of research on conceptual change* (pp. 61–82). New York: Routledge.

Damasio, A. (1994). *Descartes's error: Emotion, reason and the human brain*. New York: Avon Books.

Gibbs, R. W. (2006). *Embodiment and cognitive science*. New York: Cambridge University Press.

Glenberg, A. (1997). What memory is for? *Behavioral and Brain Sciences, 20*, 1–55.

Gómez Crespo, M. A. (2008). *Aprendizaje e instrucción en química*. El cambio de las representaciones de los estudiantes sobre la materia. Madrid, Spain: CIDE/MEPSD.

Gómez Crespo, M. A., & Pozo, J. I. (2004). From everyday to scientific knowledge: Understanding how matter changes. *International Journal of Science Education, 26*(11), 1325–1343.

Karmiloff-Smith, A. (1992). *Beyond modularity: A developmental perspective on cognitive science*. Cambridge, MA: MIT Press/Bradford Books.

Meheut, M. (2004). Designing and validating two teaching-learning sequences about particle models. *International Journal of Science Education, 26*(5), 605–618.

Pozo, J. I., & Gómez Crespo, M. A. (2005). The embodied nature of implicit theories: The consistency of ideas about the nature of matter. *Cognition and Instruction, 23*(3), 351–387.

Pozo, J. I., Gómez Crespo, M. A., & Sanz, A. (1999). When change does not mean replacement: Different representations for different contexts. In W. Schnotz, S. Vosniadou, & M. Carretero (Eds.), *New perspectives on conceptual change* (pp. 161–174). Oxford, UK: Elsevier.

Spelke, E. (1994). Initial knowledge: Six suggestions. *Cognition, 50*, 431–445.

Stavy, R. (1995). Conceptual development of basic ideas in chemistry. In S. M. Glynn & R. Duit (Eds.), *Learning science in schools* (pp. 131–154). Mahwah, NJ: Lawrence Erlbaum Associates.

Vosniadou, S., Vamvakoussi, X., & Skopeliti, I. (2008). The framework theory approach to the problem of conceptual change. In S. Vosniadou (Ed.), *International handbook of research on conceptual change* (pp. 1–34). New York: Routledge.

Williamson, V., Huffman, J., & Peck, L. (2004). Testing students' use of the particulate theory. *Journal of Chemical Education, 81*(6), 891–896.

CHAPTER 6

"I Thought the Smaller the Wave Was, the Louder the Sound Was"

Highlighting Negotiation in Developing Meta-Representational Competence

Christopher G. Wright

INTRODUCTION

This chapter highlights the role of *negotiation* in exploring children's meta-representational competence (MRC) (diSessa, 2004; diSessa & Sherin, 2000). The framework of MRC focuses on children's abilities "to select, produce, and productively use representations, but also the [students'] abilities to critique and modify representations and even to design completely new representations" (diSessa & Sherin, 2000, p. 386). Specifically, the focus is on the representational practices exhibited within a group of middle school, African American boys. Until recently, studies within the MRC framework have primarily focused on identifying the range of competencies demonstrated by students, including those competencies utilized during the production of representations (constructive resources) and those competencies utilized during the evaluation of representations (critical resources). The focus of the investigation is exploring students' constructive resources that include, but are not limited to, students' competencies and experiences with drawing (Azevedo, 2000; Sherin, 2000), competencies in utilizing color in drawings (Azevedo, 2000), and competencies with manipulating line segments in drawings (Sherin, 2000). More recent studies involving MRC have expanded methodological approaches to also emphasize and examine the practice of representation and the development of criteria (diSessa, 2002) for a "good" representation as being "socially mediated, open to change, and negotiated within ongoing activity" (Danish & Enyedy, 2007, p. 6). Combining these two methodological approaches, below I highlight the negotiation process, both internal and external, in which the boys engage during an attempt to construct a shared meaning for their representational understanding (Gutiérrez, 2002; Nasir, Rosebery, Warren, & Lee, 2006; Rogoff, 2003). Using this as a foundation, this chapter specifically focuses on children's abilities in making use of various features of a line segment (Sherin, 2000) when representing sound transmission. Thus, I demonstrate the competencies that a group of children bring to the practice of co-constructing meanings for representations, as well as examine the complexities of engaging in this local practice of producing, using, and negotiating representations of sound transmission.

Emphasizing the examination of external representations, this chapter calls attention to Enyedy's (2005) definition of representation as the "act of highlighting aspects of our experiences and communicating them to others and ourselves" (p. 427). This conceptualization highlights the social element of representation that is the primary component of this study. In addition, Enyedy's (2005) conceptualization recognizes that meanings are continuously developed and interrogated within a group of individuals, as well as within the individual, as they interact with external representations. This conceptualization of representation as a practice made sense for this study, as I examined how a group of boys negotiated their varied experiences with and understandings of sound transmission, as represented on paper. As the boys produced and utilized representations that focused on a specific aspect of sound transmission, I conjectured that various components of their understanding, how to represent those understandings, and the intricacies of the negotiation process used to explore the science and the representations of the science would become explicit and further refined.

My focus on MRC is informed by my experience of how external representations are typically employed in science classrooms. Children are frequently asked to engage with scientific ideas through images, ranging from drawings and maps to graphs and diagrams. For instance, children in science classrooms frequently encounter diagrams of the water cycle, free-body diagrams, orthographic projection drawings, and images of the engineering design process (Bar, 1989; Wendell & Lee, 2010). Unfortunately, children's engagement with visual images in science is often structured by the assumption that these images are self-explanatory and assist in making the phenomena easier to understand (Lowe, 2000). This assumption supports the idea that visual images, especially those found in disciplines that are supposedly objective in nature (e.g., science, engineering, and mathematics), have inherent meanings that students are expected to comprehend and interpret with appropriate instructional support. This assumption contradicts the underlying theoretical position informing my work with the boys, as well as studies within an MRC framework (diSessa, 2004; diSessa & Sherin, 2000), that individuals develop ways of conceptualizing, representing, evaluating, and engaging the world as they negotiate varied repertoires of practice across domains of experiences (Gutiérrez & Rogoff, 2003; Lee, 2003; Nasir, 2000; Nasir et al., 2006). The belief that scientific images possess inherent meanings ignores the real-life experiences and prior knowledge that students bring to the science classroom and to the understanding of these images (Kress & van Leeuwen, 1996; Lemke, 1989; Roth, Pozzer-Ardenghi, & Han, 2005), thus continuing to marginalize a large number of children from fully connecting with and participating in school science (Warren, Ogonowski, & Pothier, 2005). In order to highlight and privilege the boys' life experiences and prior knowledge with sound transmission, I also focus on another principle of MRC, the production of external representations, in addition to the focus on their process of negotiating a shared meaning for the produced representations.

The central idea driving this research is the exploration of the varied competencies that children bring to the production and use of external representations. Ultimately, the goal would be to further nurture and utilize these competencies as tools for teaching and learning in urban science classrooms. To explore a group of students' competencies and complexities with developing a shared meaning for representational features, I will first examine meta-representational competence and specifically highlight one constructive resource described as students' ability to associate meaning to features of a line segment. Next, the context for the study's design and methodology will be described. This chapter will utilize individual critique episodes of four drawings produced within the group of boys to highlight various competencies demonstrated by the group and the multiple complexities that arose during discussions. Lastly, the implications that the results could have for teaching and learning science within urban classrooms will be considered.

EXPLORING META-REPRESENTATIONAL COMPETENCE

Studies utilizing the MRC framework have a common interest of exploring the comprehensive range of competencies that children bring to the task of producing and using external representations (diSessa, 2004; diSessa & Sherin, 2000). This range of competencies includes children's abilities to select, produce, use, critique, and modify representations. MRC studies are deeply rooted in the premise that children's intuitive ideas and previous experiences serve as the foundation for developing deeper understandings of the practice of representation, as well as the mathematical and scientific concepts being represented.

Previous work in MRC has looked beyond what has been considered children's "misconceptions" in producing and using external representations. MRC has privileged children's ways of knowing, thinking, and conceptualizing as precursors to deeper representational and scientific understanding. For example, *misconceptions* literature regarding representations has primarily focused on children's mistakes and/or inabilities to produce and use specific types of representations, including graphs and tables (Leinhardt, Zaslavsky, & Stein, 1990). In contrast, diSessa, Hammer, Sherin, and Kolpakowski (1991) argue that through a process of iterative design that included interactions and critiques with peers, a group of 6th-grade students produced representations of motion that resembled conventional graphs of speed versus time. Although these students had not received any direct or formal instruction in designing representations and had not previously produced conventional representations of graphs, this particular group of children drew on their intuitive understandings while refining their representations through a process of production and interpretation. My contribution to this line of inquiry features the social element in a group's developing competency with representing sound transmission. Adhering to MRC's general premise that children bring a range of competencies to the practice of representation, I emphasize how these competencies are continuously negotiated and explored within a social setting while attempting to develop a shared meaning.

Other research has also explored children's MRC in regard to various scientific phenomena, including the representation of height in two-dimensional maps (Enyedy, 2005), the representation of model and real landscapes (Azevedo, 2000), and further explorations into the concept of depicting motion through drawings (Sherin, 2000). These studies, building from diSessa et al. (1991), have identified a range of *constructive resources* that contribute to children's overall competency in the practice of representation. *Constructive resources* can be described as the resources that children bring to the task of "inventing" representations that, as explained by Azevedo (2000), "derive from a combination of learned abilities, conventions, and rules" (p. 445). Thus, *constructive resources* are those core capabilities and elements of children's prior knowledge that contribute to their competencies in producing and using external representations. For instance, both Azevedo (2000) and Sherin (2000) recognize children's prior experiences with the practice of *drawing*. As a constructive resource, *drawing* focuses on and explores the set of conventions and techniques that children have developed for representing aspects of the world on paper (Olson, 1994).

Children's Knowledge with Manipulating Line Segments

During explorations that involved children's representations of a moving vehicle, Sherin (2000) recognized children's competencies with manipulating and making use of various features of line segments. As a constructive resource, or a resource that contributed to children's abilities to invent representations, children's abilities in manipulating line segments included the isolation and altering of various features in order to convey specific meanings. Students were found to manipulate features such as height, slope, length, thickness, and/or color of lone segments. For example, students were observed manipulating the height of line segments in order to depict the changing speed of the moving vehicle. To represent a car going from a high to a slower speed, students incorporated taller line segments that gradually decreased in height as a person read the drawing from left to right. Conversely, representing a car going from a slower speed to a faster speed was illustrated by using shorter lines that gradually increased in height.

In another example, students also utilized their knowledge for manipulating the slope of line segments in order to represent a vehicle's changing speed (Sherin, 2000). Here, vertical lines depicted a vehicle's stationary position, while horizontal lines depicted its maximum speed. To represent the vehicle's gradual speed, students gradually increased the slope of the line segments within their drawing. Thus, students' knowledge and competencies with manipulating features of line segments to highlight various meanings were recognized as a resource in children's inventions of representations of motion. Building upon the recognition of this resource, or competency, this chapter explores how a particular group of boys negotiated their associations of varied meanings of the same feature, the length of a line segment. Despite not illustrating a final shared meaning within the group of boys, this chapter highlights the complexities and processes encountered while attempting to develop a shared understanding.

CONTEXT OF THE WORK

In this chapter, I take a close look at the representational experiences of five African American boys, Earl, Floyd, Isaac, Kenneth, and Tim, as they participate in a 4-day exploration of sound transmission. The goal of this exploration was to examine children's ideas regarding sound transmission and their competencies in representing these ideas through drawings. The exploration sessions were conducted on consecutive days during a nonscheduled vacation (i.e., the boys were not in school due to a huge snowstorm in the area). Employing a discourse analysis methodology (Gee, 1999), I utilized students' talk regarding their drawings as a source of data in determining their dynamic and in-the-moment processing of the scientific phenomena. Additionally, student talk was combined with an analysis of students' drawings (Azevedo, 2000; diSessa, 2004; diSessa et al., 1991) in order to further develop an understanding of their representational practices and understandings. This exploration was not designed as a formal intervention intended to "teach" the boys about the science of sound, nor to specifically alter or adjust their ideas of sound transmission. On the contrary, the exploration was designed to elicit and document how the boys' constructive resources supported their efforts in co-constructing meanings of sound transmission (Linder, 1992).

A major component in the design and analysis of the activities within the 4-day exploration were critique episodes, where the boys engaged in the practice of analyzing and critiquing one another's drawings that were produced during the exploration. This chapter specifically examines several opportunities for critique during Session 3 of the sound transmission exploration. The following analysis will emphasize the students' ideas about the change in volume over distance and their ways of representing this idea. The idea of change in volume over distance is used to describe the process through which sound waves, emanating from a source of sound, become less intense as they move out spherically from the sound producer (Parker, 2009).

HIGHLIGHTING NEGOTIATION WITHIN THE GROUP

The following results are presented in the form of case studies of four drawings produced during Session 3.

The Group Critique of Earl's Drawing

Session 3 was designed to provide the boys with shared experiences with sound transmission and to provide further opportunities to analyze these experiences through drawings and critique. Initially, the boys closed their eyes and silently listened to the surrounding sounds for approximately 2–3 minutes. Following this experience, they discussed descriptions of the sounds that each person heard. Example descriptions included the source of the various sounds, the intensity of the various sounds, the duration for which each sound lasted, and the various pitches heard by the group. Following this discussion period, each boy produced a drawing that served as an external representation of the sounds they experienced. To introduce the group's complexities in co-constructing meanings to be associated with the length of a line segment, I initially highlight Earl's first drawing from Session 3 (see Figure 6.1) and the brief critique period that followed (see Excerpt 1).

Figure 6.1. Earl's First Drawing Produced During Session 3

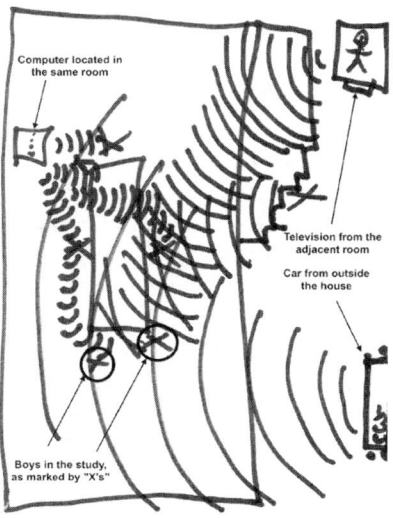

Excerpt 1. The Group Critique of Earl's First Drawing

1. *Kenneth*: It's [the sound coming from the car's horn] very loud.
2. *Floyd*: It [the sound waves from the car's horn] went the furthest.
3. *Chris*: It did what?
4. *Floyd*: It went the furthest.
5. *Chris*: How do you; why do you say that?
6. *Floyd*: 'Cause it [the sound waves from the car's horn] got through the walls and everything else is closer. And that one [the sound waves from the car's horn] is the furthest. It [the sound waves from the car's horn] got the biggest wavelength.
7. *Chris*: What do you mean by wavelength?
8. *Floyd*: Everything else, their waves are kinda small compared to the horn.
9. *Chris*: And what did you say, Kenneth?
10. *Kenneth*: Um, I put the horn.
11. *Chris*: No, I mean about this [redirecting Kenneth's attention to Earl's drawing] here. What did you say about that?
12. *Kenneth*: This [the car's horn] would be the loudest because it has the biggest sound waves and it traveled the furthest.

[A period of 2 minutes passes in the conversation where the boys discuss the idea of duration.]

13. *Chris*: Okay, um, are there ways that you can show that the horn was loud //[1]
14. *Tim*: // Yes! [Pointing to a feature on his drawing.] //
15. *Chris*: // But only lasted like for two seconds as opposed to the television not being as loud, but went the entire time? Are there ways that you can show that?
16. *Floyd*: I think you can make the; it's the size of the waves.
17. *Chris*: You think it's the size of the waves?
18. *Floyd*: Like, if you was going, um. How we [pointing to Kenneth] said that the smaller waves, that they're the loudest, but then as they get bigger, that's how they kinda get lower [volume], and you can make; if it was the loudest, but for a short period of time, you can make a lot of little short ones.

I first call attention to Kenneth's initial description of the car horn's sound waves (see line 1). His association of a louder volume with the representation of longer waves is important to highlight for several reasons. First, throughout the previous 2 days of the sound exploration, Kenneth consistently maintained the idea that a louder volume is typically represented by shorter waves. Kenneth's initial interpretation of Earl's drawing contradicts his previous associations. Next, Kenneth's assertion of "it's very loud" provided the group with a foundational association that can now be interrogated and negotiated within the group as they critique Earl's drawing, as well as the drawings that follow. Lastly, Kenneth's initial assertion, displayed in line 1, will be important in highlighting the idea that complexities and tensions did not only exist between group members, but also existed within individuals. Later in this chapter, Kenneth will associate an alternate meaning with longer waves that will challenge the group to further consider the development of a consensus meaning for the length of a wave.

Next, I skip to the meaning that Floyd attached to a longer wavelength (see line 6), as he, too, initially associated this feature with a louder volume. Here, Floyd is agreeing with Kenneth's assertion (see line 1 and line 12) that the longer waves in Earl's drawing (see Figure 6.1) represent a louder volume. Again, this detail is important because Floyd, too, consistently utilized shorter waves to represent louder volumes during Sessions 1 and 2. This group tension regarding the meaning of the length of a wave as it relates to volume, as well as the individual tension, is displayed later in this chapter as the group continued to discuss Earl's drawing.

When provided with the question of how you can represent the loudness of the horn (see line 13), Floyd (see line 18) offered the suggestion of utilizing shorter waves to represent the loud volume. Here, Floyd has clearly contradicted his (see line 6) and Kenneth's (see line 1 and line 12) earlier interpretation of Earl's drawing, thus highlighting the complexity in co-constructing a group meaning. As I further examine this apparent tension in Floyd's interpretations of the drawings, I recognize an important detail in the tasks that were presented to Floyd and Kenneth. Initially, the two boys were asked to interpret, or read, what they believed Earl was communicating in his drawing. Floyd's later comment (see line 18) was provided when I asked the group how they would represent a louder volume. Here, Floyd initially emphasizes that he and Kenneth previously established that the "smaller waves" are the loudest. Next, Floyd describes that as the wavelength gets longer, the volume is decreasing. This small detail here is important because I maintain that Floyd and Kenneth demonstrate a competency that has not been previously identified in this chapter. Both of the boys recognized that Earl's representation could be "correct," but it was simply not the way in which they would represent a loud volume. As described by diSessa (2004), Floyd's attitude toward the meanings that Earl associated with longer waves displayed Floyd's "relatively high degree of metarepresentational sophistication" (p. 302) because he was able to understand that Earl's representation was not necessarily "wrong," but one that Floyd himself did not feel comfortable utilizing. To further illustrate the boys' competencies, as well as the complexities in co-constructing meaning, I next analyze the group critique of two other drawings in Session 3.

The Group Critique of Kenneth's and Floyd's Drawings

In the next two critique episodes to be analyzed, the boys were given the task of comparing and contrasting the volume of two separate sounds and the duration that each sound lasted through their drawings. The first sound was generated by a television in an adjacent room within the house and the other sound was generated by a car's horn located directly outside of the house. Again, this analysis will focus on the group's discussion regarding Kenneth's and Floyd's representation of volume. The critique episode (see Excerpt 2) surrounding Kenneth's drawing (see Figure 6.2) will be analyzed first.

Excerpt 2. The Group Critique of Kenneth's Second Drawing

1. *Chris*: On this drawing here, right? So, we know that the horn was louder [than the television], right? So, how is Kenneth here showing that the horn is loudest?
2. *Kenneth*: Because, even though this is not; it [the sound waves from the car horn] wasn't the longest one [in regard to duration], the waves stay short. So, it [the sound waves from the car horn] kept it loud and the volume of the horn didn't change. So, it [the volume of sound from the car horn] stayed. //
3. *Chris*: So, the short waves mean a loud sound?
4. *Tim*: That's what mine meant.
5. *Kenneth*: Yeah, a louder sound. And, um.

   [A period of 3 minutes passes in the conversation where the boys discuss the idea of duration.]

6. *Chris*: Okay, what things does anybody notice about Kenneth's drawing here? Yeah, Earl?
7. *Earl*: His television was, is like; I thought that the car waves could have been bigger. And they [waves emanating from the television] could have been a lil' bit smaller.
8. *Kenneth*: They are smaller.
9. *Chris*: Which ones?
10. *Earl*: The car waves. They [the waves from the car horn] could have been bigger and they [the waves from the television] could have been a bit smaller, but that's probably how he was hearing it.
11. *Isaac*: I notice that his waves from the car stayed a little bit constant.
12. *Chris*: They stayed what? Oh, constant, you said? Okay.
13. *Isaac*: From that point, they stayed constant.

14. *Floyd*: I noticed that the television waves took up most of the room than the horn [can be interpreted as the television waves took up more room than the waves from the car horn].

15. *Chris*: Oh, okay, and what does that mean?

16. *Floyd*: It means that the; I guess it means that the television was louder in this room than the; you could hear the television more in this room than the car.

17. *Chris*: And what does the waves staying constant mean to you?

18. *Isaac*: Well, I thought it meant that the; that the sound stayed. Like, if you were closer to it, to the car, then it'd be the same as if you were in here [in the room].

I feature the critique of this drawing in order to highlight the tensions that I recognized as the group continued to associate different meanings with the feature of a line segment's length. First, I call attention to Kenneth's (see line 2) description of his drawing. Here, Kenneth explicitly stated that "the waves stay short, so it kept it loud." The meaning that Kenneth associated with the shorter waves was a loud volume. This association was more consistent with Kenneth's previous drawings and descriptions provided during Sessions 1 and 2, but contradicted with his earlier interpretation of Earl's first drawing (see Excerpt 1, line 1 and line 12). In contrast, Earl (see line 7) appeared to be suggesting the opposite meaning to what Kenneth intended, thus drawing attention to the tension that pervaded the group throughout the four sessions. In this instance, the group had already agreed

Figure 6.2. Kenneth's Second Drawing During Session 3

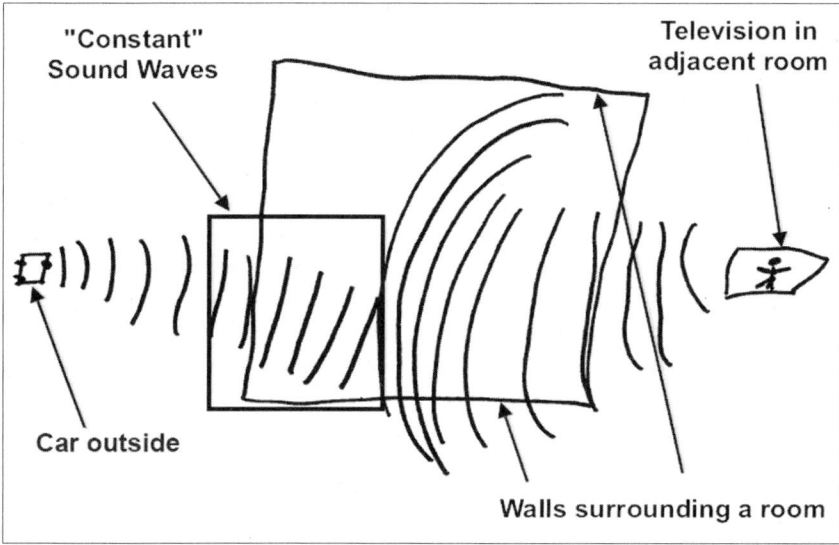

that the car horn would have a louder volume than the television playing in the adjacent room. Here, I believe Earl associated the meaning of longer waves with a louder sound, as he offered the suggestion, "the car waves could have been bigger." Throughout the four sessions, there were several boys, including Floyd and Kenneth, who consistently associated short waves with a loud volume and other boys, including Earl and Isaac, who held the opposite association.

Additionally, I single out Earl's (see line 10) suggestion regarding Kenneth's representation. Here, Earl suggested that because Kenneth was hearing, or experienced, the television's volume as louder, this would explain why the television waves were longer in his drawing than Earl himself would have drawn them. This statement is interesting to note when we think about Enyedy's (2005) definition of representation as "the act of highlighting aspects of our experience and communicating them to others and ourselves" (p. 427). Was Earl suggesting that although he did not necessarily agree with Kenneth's drawing, he understood that Kenneth was merely "highlighting an aspect of his experience"? Although I cannot definitively answer this question, I do suggest that Earl does display a level of meta-representational sophistication (diSessa, 2004), as he was able to implicitly "accept" Kenneth's drawing because it was "probably how he was hearing it" (see line 10).

As the critique continued, Floyd (see line 16) also read Kenneth's drawing as depicting that the television was producing a louder volume than the car horn. Again, as earlier noted, Floyd consistently associated longer waves with quieter volumes, so it was interesting to notice his interpretation of Kenneth's drawing because it contradicted the associations that he consistently had during the 4-day exploration. The interaction between Kenneth, Earl, and Floyd brings attention to the underlying friction in the co-construction of a shared meaning for representing volume by manipulating the length of a line segment.

Lastly, Kenneth's drawing also offered the group an opportunity to expand their exploration into the meanings they associated with the length of a line segment. In this instance, the ways in which the length of the line segment changed over distance were highlighted during the critique. Kenneth's drawing prompted the boys to introduce a new idea: the idea of "constant" waves (see line 11 and line 18). Here, Isaac focused on how the lines did not change in length, thus reading Kenneth's drawing as communicating that the volume of the car's horn remained constant. Although the group did not further examine this meaning of "constant waves," it does highlight another aspect of this feature of associating meaning to the length of a line segment. In this case, the reading of the representation was not simply isolated to the length of one line segment, but was examined through a focus on the relationships among several line segments and the change in length between line segments.

To further complicate Floyd's previous readings of Earl (see Figure 6.1) and Kenneth's (see Figure 6.2) drawings, I now use Floyd's description (see Excerpt 3) of his own drawing (see Figure 6.3).

Figure 6.3. Floyd's Second Drawing Produced in Session 3

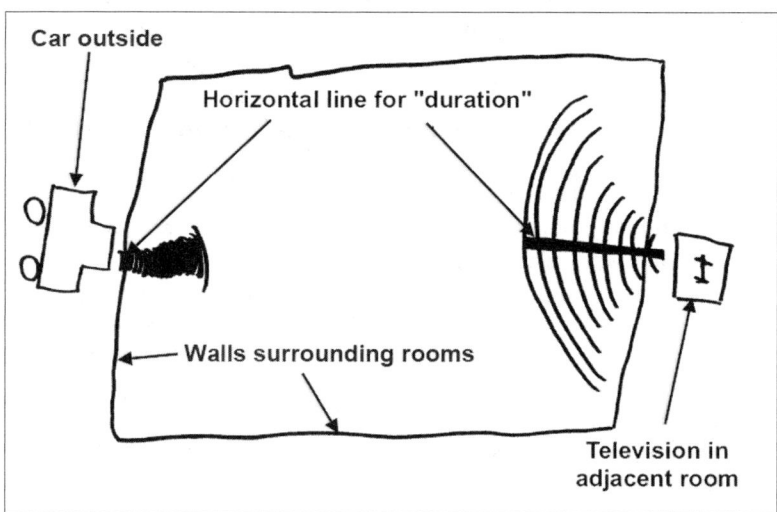

Excerpt 3. Floyd's Description of His Second Drawing

1. *Chris*: So, what's happening here?
2. *Floyd*: I showed. So, what you want me to say?
3. *Chris*: How are you showing that the horn's louder?
4. *Floyd*: The horn, I showed. In my drawing, I showed that the horn was louder because the waves are smaller. And, I said that the waves are smaller means that they are packed more behind them and you might can't see it but, like this one [the horizontal line within the sound waves from the car horn] I drew a line and it showed how long it was [regarding the duration that the sound lasted]. How long the sound was over the course of 2 minutes. That was for the car. The television, the waves like, they spread out showing how they fade because if you was in the living room, you would hear it better than you would hear it now [in the room in which the boys were in]. The line [the horizontal line within the sound waves from the television] is longer because it [the sound] lasted the whole 2 minutes of the time.

Floyd's drawing (see Figure 6.3) and subsequent verbal statements are highlighted here to further exemplify both the competencies and the complexities that existed in producing and using representations within a social context. Floyd's description of his drawing (see line 4) emphasizes his consistent use of shorter waves to represent a louder volume throughout the four sound exploration sessions. Floyd's ability to keep his understanding and use of this feature constant, while also being able to interpret and understand how others may contradict this, is further evidence of his meta-representational sophistication (diSessa, 2004). In addition, it also highlights the difficulties with coming to a group consensus on the use of this feature.

Analyzing a "Professional" Representation of Sound

In this final episode, I look to further draw out the boys' competencies with constructing and developing representations of sound transmission while at the same time I continue to explore the process of developing a group meaning that is associated with the length of line segments. A small portion of the 4-day sound exploration consisted of the examination of how professionals, in this case professional architects, represent sound transmission. Session 3 concluded with one of these opportunities, in which the group was provided a sample, nonconventional architectural site map that detailed the intensity of various sounds surrounding the proposed site (see Figure 6.4). In this drawing, seven "sound waves" (each is individually numbered) are used to represent seven different sounds and their varying volumes. When this drawing was put in front of them, the group was presented with the task of simply communicating things they noticed within the drawing. Excerpt 4 details a brief moment during the critique that highlights the possibility that the architect utilized the feature of length of line segments in the drawing.

Figure 6.4. The Architectural Site Map Analyzed by the Group During Session 3 (Generated by the Researcher, but Informed by White, 2004)

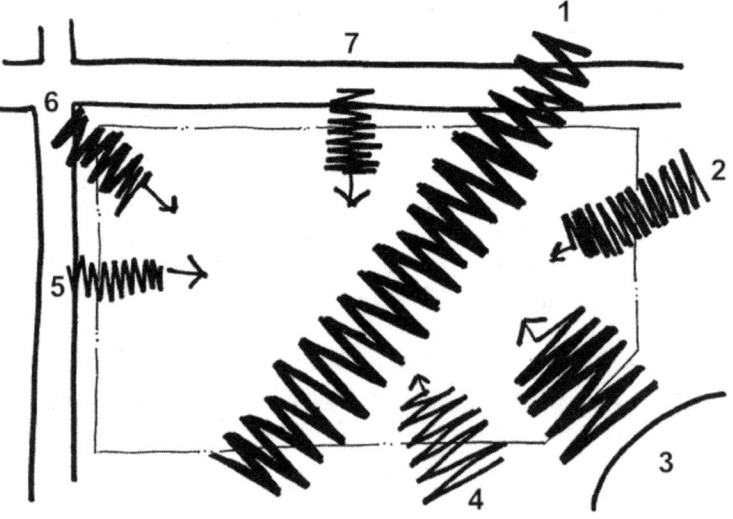

Excerpt 4. The Group's Analysis of a Site Map

1. *Earl*: I think they make it like, how this one (sound wave #5) is like small. It is then they made like the sound. This (sound wave #1) the biggest and then it's that one (sound waves #3). This (sound wave #1) and this, the least (sound wave #5).
2. *Chris*: Why do you say that?
3. *Earl*: 'Cause like how big it (sound wave #1) is, like. I would go by the sizes of; by the size of it (the sound waves).
4. *Chris*: Tim?
5. *Tim*: I notice that, ah. They show how the sound waves, ah. By, what's that called?
6. *Chris*: Those squiggly line things?
7. *Tim*: Yeah, by making more zigzags than the one. Hold up. Making, they made more zigzags for the one that's louder (sound wave #1), and made less for the one that was not that big a sound (sound wave #5).
8. *Chris*: Okay, Kenneth, you were about to say something?
9. *Kenneth*: Yeah, Earl said something about how big or how small they was but, all right, that one is smaller (sound wave #5) but that one is bigger (sound wave #1). I thought the smaller the wave was, the louder the sound was.

In this exchange, Earl (see line 1) initially identifies the "biggest" sound wave as having the loudest volume, a stance that he has maintained throughout the group's 4-day exploration into sound. Then, Kenneth (see line 9) disagrees with Earl's statement by stating that he thought that they said the smaller sound waves would represent a louder volume. First, this tension of developing a shared meaning associated with longer and/or shorter waves remained in play throughout the 4 days of the exploration. Next, despite the fact that the "professional" utilized a different notation (i.e., "zigzags" versus the boys' use of crescents), the boys still pulled from their earlier experience with producing and using representations of sound to make sense of this site map. I contend that this illustrates a continuing development and adjustment in the resources that the boys bring to their science exploration. These resources are not static ideas or approaches that remain the same throughout their exploration, but are dynamic in nature and always being fine-tuned.

THINKING ABOUT THE ROLE OF REPRESENTATIONS IN STEM CLASSROOMS

This analysis of four individual critique episodes offered a perspective on the competencies that the group demonstrated regarding the production and use of representations of scientific ideas. Additionally, this analysis recognized some of the complexities and tensions that the group faced as they attempted to co-construct the meanings associated with their use of longer and/or shorter sound waves. The meanings of these representations were not inherent, despite the producer's intent, but, rather, continuously interrogated and developed through group interaction and discussion. Sherin (2000) stated that in order for representations to be successful and to have the possibility of developing into conventional use, children must maintain a consistent meaning for the usage of the incorporated representational elements (i.e., features of a line segment). In this chapter, the group expressed various, and plausible, ways for using and interpreting the feature of the length of a line segment. Using Sherin's (2000) point as a foundation, this chapter explored and highlighted the complexities in the developing conventional representations. The development of conventional representations is not a straightforward task, but a task that is socially mediated and constructed over time and through a variety of communities.

The use of increasing or decreasing length of the line segments carried multiple and dynamic meanings within the group as they represented the idea of change in volume over distance. The variation in meaning highlights the intricacies that persisted within the group in developing a shared understanding of this feature. In addition, I argue that the production and critique of representations was very useful in providing the boys with opportunities to extract, refine, and construct meaning within the context, thus leading to a call for more research of this kind of activity within science classrooms. Even though the drawings incorporated the same representational notation, the crescent, there still was limited consensus regarding the crescents' meaning by the end of the fourth exploration session. This limited consensus does not emphasize limited competencies with developing or using representations within the group. On the contrary, the analysis presented here showcases how representations are dependent upon the perceptions of those who are interpreting their functions and the contexts within which the representations and associated elements are being used. This idea is important as we begin to think about the role of negotiation in the general development of representations and the ways in which people come to understand the possible various meanings.

By highlighting the practice of critique and negotiation within this chapter, I advocate for the need to further explore the potential roles that representations can play in teaching and learning within STEM classrooms. This work describes how drawings created by a group of boys were positioned as "objects of inquiry" (Warren & Rosebery, 2011). Here, the boys continuously developed their understanding of sound transmission by examining, unpacking, and interrogating the multiple meanings associated with longer and/or shorter sound waves. In these instances of critique and negotiation, the group explicitly explored how meanings could change, depending upon the context in which the representation was created and/or interpreted. This competency with creating and interpreting representations within varied contexts could even be extended to representations that are not created by children. Typical representations within STEM classrooms, such as graphs, diagrams, and charts, also have the potential to be transformed into "objects of inquiry." Classroom practice could include opportunities for children to explore and examine the varied meanings within these representations, thus foregrounding the role of the social construction of knowledge within science.

NOTES

1. // is used to show an interruption in conversation by another speaker.

REFERENCES

Azevedo, F. (2000). Designing representations of terrain: A study in metarepresentational competence. *Journal of Mathematical Behavior, 19*, 443–480.

Bar, V. (1989). Children's views about the water cycle. *Science Education, 73*(4), 481–500.

Danish, J., & Enyedy, N. (2007). Negotiated representational mediators: How young children decide what to include in their science representations. *Science Education, 91*(1), 1–35.

diSessa, A. (2002). Students' criteria for representational adequacy. In K. Gravemeijer, R. Lehrer, B. van Oers, & L. Verschaffel (Eds.), *Symbolizing, modeling, and tool use in mathematics education* (pp. 102–129). Dordrecht, Netherlands: Kluwer Academic Publishers.

diSessa, A. (2004). Metarepresentation: Native competence and targets for instruction. *Cognition and Instruction, 22*(3), 293–331.

diSessa, A., Hammer, D., Sherin, B., & Kolpakowski, T. (1991). Inventing graphing: Metarepresentational expertise in children. *Journal of Mathematical Behavior, 10*, 117–160.

diSessa, A., & Sherin, B. (2000). Metarepresentation: An introduction. *Journal of Mathematical Behavior, 19*, 385–398.

Enyedy, N. (2005). Inventing mapping: Creating cultural forms to solve collective problems. *Cognition and Instruction, 22*(3), 293–331.

Gee, J. P. (1999). *An introduction to discourse analysis: Theory and method.* London, UK: Routledge.

Gutiérrez, K. (2002). Studying cultural practice in urban learning communities. *Human Development, 45*, 312–321.

Gutiérrez, K., & Rogoff, B. (2003). Cultural ways of learning: Individual traits or repertoires of practice. *Educational Researcher, 32*(5), 19–25.

Kress, G., & van Leeuwen, T. (1996). *Reading images: The grammar of visual design.* Abingdon, UK: Routledge.

Lee, C. D. (2003). Why we need to re-think race and ethnicity in educational research. *Educational Researcher, 32*(5), 3–5.

Leinhardt, G., Zaslavsky, O., & Stein, M. (1990). Functions, graphs, and graphing: Tasks, learning, and teaching. *Review of Educational Research, 60*(1), 37–42.

Lemke, J. (1998). Multiplying meaning: Visual and verbal semiotics in scientific text. In J. R. Martin & R. Veel (Eds.), *Reading science: Critical and functional perspectives on discourse of science* (pp. 87–113). London, UK: Routledge.

Linder, C. J. (1992). Understanding sound: So what is the problem. *Physics Education, 27,* 258–264.

Lowe, R. (2000). Visual literacy in science and technology education. *UNESCO International Science, Technology, & Environmental Education Newsletter, XXV*(2), 1–2.

Nasir, N. (2000). "Points ain't everything": Emergent goals and average and percent understandings in the play of basketball among African American students. *Anthropology & Education Quarterly, 31*(3), 283–305.

Nasir, N., Rosebery, A., Warren, B., & Lee, C. D. (2006). Learning as a cultural process: Achieving equity through diversity. In R. K. Sawyer (Ed.), *The Cambridge handbook of the learning sciences* (pp. 489–504). New York: Cambridge University Press.

Olson, D. R. (1994). *The world on paper: The conceptual and cognitive implications of writing and reading.* Cambridge, UK: Cambridge University Press.

Parker, B. (2009). *Good vibrations: The physics of music.* Baltimore, MD: The Johns Hopkins University Press.

Rogoff, B. (2003). *The cultural nature of human development.* New York: Oxford University Press.

Roth, W. M., Pozzer-Ardenghi, L., & Han, J. Y. (2005). *Critical graphicacy: Understanding visual representation practices in school science.* Dordrecht, Netherlands: Springer.

Sherin, B. (2000). How students invent representations of motion: A genetic account. *Journal of Mathematical Behavior, 19,* 399–441.

Warren, B., Ogonowski, M., & Pothier, S. (2005). "Everyday" and "scientific": Rethinking dichotomies in modes of thinking in science learning. In R. Nemirovsky, A. Rosebery, J. Solomon, & B. Warren (Eds.), *Everyday matters in science and mathematics: Studies of complex classroom events* (pp. 119–148). Mahwah, NJ: Lawrence Erlbaum Associates.

Warren, B., & Rosebery, A. (2011). Navigating interculturality: African American male students and the science classroom. *Journal of African American Males in Education, 2*(1), 98–115.

Wendell, K. B., & Lee, H. S. (2010). Elementary students' learning of materials science practices through instruction based on engineering design tasks. *Journal of Science Education and Technology, 19*(6), 580–601.

White, E. (2004). *Site analysis: Diagramming information for architectural design.* Tallahassee, FL: Architectural Media.

Representational Competence in Science Education

Its Fundamental Role in Today's Science Teaching

Tina A. Grotzer

Making meaning is inherent to the purpose of science. Our quest for explanation invites us to consider how and why our world works the way it does. These understandings are not merely academic—they are critical to our sense of well-being. They enable us to develop expectations and a sense of control because they allow us to intervene as we attempt to steer future outcomes. Philosophical discussions aside, they engender a notion of free will and imbue societies and individuals with a sense of purpose.

The process of how we make meaning in science and how it leads to warrantable conclusions has evolved over time. It has shifted with our access to new and different tools and paradigms. From Aristotle and the Church to Kuhnian notions of socially constructed paradigms, the fundamental features of making meaning have also changed (Bauer, 1992; Chalmers, 1999). Earlier notions that facts accumulate by disproving bold, falsifiable statements (e.g., Popper, 1934/2002) have given way to conceptions of meaning driven by the framing and constraints of particular, socially negotiated paradigms (Kuhn, 1962). New technologies developed in fields such as synthetic biology and gene sequencing invite new modes of meaning-making (Liu & Grotzer, 2009).

Paralleling these shifts in the epistemology of science have been considerable shifts in how we view the process of meaning-making in science education. These shifts have impetus well beyond the study of the epistemology of science. The nascent fields of cognitive science and the learning sciences lend insight into the human mind and the nature of learning dispelling the notion that we construct understanding of our world by accumulating knowledge. The rich literature on students' sense-making of particular concepts grounds these shifts (e.g., Driver, Guesne, & Tiberghien, 1985), while theories on why students hold the particular conceptions that they do (e.g., diSessa, 2006) frame the instructional implications.

The chapters in this section contribute to our revised conceptions of science and what it means to be scientifically literate. They underscore three essential features of actualizing these new visions: (1) The importance of students' meta-representational skills in enabling and revealing their meaning-making; (2) a focus on individual cognitive architecture—that it prioritizes certain patterns of perception and modes of interpretation that we must be cognizant of; and (3) the socially mediated construction of knowledge.

The importance of students' meta-representational skills in enabling and revealing meaning-making. The authors argue persuasively, albeit differently, for the importance of tapping into students' meaning-making and the critical role of meta-representational knowledge in doing so. A key point is that students are expected to grasp the inherent meaning in representations without having had the opportunities to develop those competencies.

Wright argues for the importance of a deep visual language in enabling students to share their sense-making. Pérez-Echeverría, Postigo, and Marín-Oller consider how the features embedded in graphical representations impact students' ability to map between the graphs and informational text. They found affordances in students' abilities and also some difficulties that might be predicted based upon the particular graphical elements. Gómez Crespo, Pozo, and Gutiérrez Julián reveal students' struggles to make sense of the microscopic given the macroscopic nature of our perceptions, and they identify difficulties in reasoning about relationships between these levels.

A focus on individual cognitive architecture. The chapters adopt slightly different views of the learner, with implications for instructional practice. Pérez-Echeverría and colleagues discuss divergences between expert and student reasoning and the importance of identifying and teaching to those. Wright focuses on building upon the generative resources students bring to their learning. He underscores how the representations created by the students are dependent upon their individual perceptions—how they interpret particular functions, associate elements, and the surrounding contexts. He discusses variations in these skills across individuals and groups and the importance of helping all students build them by bringing their resources to bear.

Gómez Crespo and colleagues seek explanations for why we generate the kinds of explanations that we do. They argue that our cognitive architecture prioritizes certain patterns of perception and modes of interpretation and that we must be cognizant of these. They discuss how our experiences and embodied knowledge create powerful sets of assumptions that influence our reasoning, such as carrying forth agent-oriented causal perceptions from the macroscopic world into the microscopic world. Instruction needs to bear in mind these powerful influences as minds interact with scientific concepts.

The socially mediated construction of knowledge. Wright emphasizes the broader sociological perspective. He argues that meta-representational skills need to be negotiated in a social context so that students develop shared ways of conceptualizing, representing, evaluating, and engaging with them. Constructing understanding within this socially mediated context is important for allowing students access to cultural conventions and for helping them see that conventions are a product of such social interaction. Without such opportunities, students are marginalized; cultural constructions vary, so we cannot assume access to the same repertoire of explanatory tools (Nasir, Rosebery, Warren, & Lee, 2006).

Wright considers how five boys communicated and negotiated meaning over 4 days using line segments in the communication of sound. He illustrates how tensions between ideas were generative in producing discussion that drove further meaning-making. He astutely questions whether the amount of time was a limiting factor; other features may have emerged in a larger group over a longer time. In my own classroom studies, I have witnessed an evolutionary process where students adopted those representational concepts that most effectively communicated ideas to others. With a larger population of concepts, through *survival of the fittest*, the most powerful conventions emerged.

Collectively, this work makes a persuasive argument for studying understanding of representational features, making meaning visible, and enabling students to learn meta-representational skills. It argues that representational knowledge must be socially negotiated in classrooms as it is in the sciences. Like recent work on argumentation and modeling (e.g., Lehrer & Schauble, 2005; Sandoval & Millwood, 2008), it invites epistemology in classrooms that more closely parallels that in the sciences.

How we reconcile the construction of socially negotiated understandings with education as it exists and societal goals for an educated citizenry is related to whether negotiated representations find a foothold in the classroom. The deep gains in epistemological knowledge must be valued over surface-level content knowledge. Helping students understand that conventions in a field are negotiated can provide eventual justification for those adopted in the disciplines.

REFERENCES

Bauer, H. (1992). *Scientific literacy and the myth of the scientific method.* Urbana, IL: University of Illinois Press.

Chalmers, A. F. (1999). *What is this thing called science?* Indianapolis, IN: Hackett Publishing Co.

diSessa, A. (2006). A history of conceptual change research: Threads and fault lines. In R. K. Sawyer (Ed.), *The Cambridge handbook of the learning sciences* (pp. 265–282). New York: Cambridge University Press.

Driver, R., Guesne, E., & Tiberghien, A. (Eds.) (1985). *Children's ideas in science.* Philadelphia, PA: Open University Press.

Kuhn, T. S. (1962). *The structure of scientific revolutions.* Chicago, IL: University of Chicago Press.

Lehrer, R., & Schauble, L. (2005). Developing modeling and argument in elementary grades. In T. A. Romberg, T. P. Carpenter, & F. Dremock (Eds.), *Understanding mathematics and science matters* (pp. 29–54). Mahwah, NJ: Lawrence Erlbaum Associates.

Liu, Y., & Grotzer, T. A. (2009). Looking forward: Teaching the nature of the science of today and tomorrow. In I. M. Saleh & M. S. Khine (Eds.), *Fostering scientific habits of mind: Pedagogical knowledge and best practices in science education* (pp. 9–36). Rotterdam, Netherlands: Sense Publishers.

Nasir, N., Rosebery, A., Warren, B., & Lee, C. D. (2006). Learning as a cultural process: Achieving equity through diversity. In R. K. Sawyer (Ed.), *The Cambridge handbook of the learning sciences* (pp. 489–504). New York: Cambridge University Press.

Popper, K. (1934/2002). *The logic of scientific discovery.* New York: Routledge.

Sandoval, W. A., & Millwood, K. A. (2008). What can argumentation tell us about epistemology? In S. Erduran & M. P. Jiménez-Aleixandre (Eds.), *Argumentation in science education: Perspectives from classroom-based research* (pp. 68–85). Dordrecht, Netherlands: Springer.

Part III

HIGHLIGHTING

CHAPTER 7

What Do Representations Represent?

Interpreting Students' Representations of Air

Tracy Noble

Students create and use a range of representations, including drawings, diagrams, gestures, and verbal analogies, to make sense of scientific phenomena. When students are discussing phenomena they cannot manipulate directly because they are too big, such as the Earth or solar system (Crowder, 1996; Vosniadou & Brewer, 1992), or too small, such as molecules (Acher & Arca, 2009; Johnson, 1998; Nussbaum, 1985; see also Chapters 5 and 9), students' representations can help researchers understand how students are thinking about these phenomena. In some cases, researchers take students' drawings or verbal descriptions of scientific phenomena as representations of some form of internal representation (e.g., a mental image or mental model) of those phenomena. What do students' representations represent? Do they represent some form of internal representation, or are they instead tools to think with, developed in the moment, in response to the task at hand and to the physical and social contexts?

This chapter argues for the second view of students' representations, through the analysis of three episodes of students' construction and use of representations of air as particles. In these episodes, students created their representations in interaction with the physical and social contexts and used them as tools to think with as they worked to explain the behavior of air in terms of air particles.

ARE REPRESENTATIONS INSIDE OR OUTSIDE?

Researchers in science education often view students' written, drawn, gestured, or spoken representations as representations of something "internal" in students' minds, such as students' mental models, conceptions, or theories regarding the phenomena under study (Adadan, Irving, & Trundle, 2009; Gentner & Gentner, 1983; Johnson, 1998; Pozo & Gómez Crespo, 2005; Vosniadou & Brewer, 1992). Such studies have often elucidated striking consistencies among the representations used by a range of students, and have provided valuable insights into students' thinking. However, the perspective taken by such studies reinforces a dualistic view of mind and world that places representations outside and mental models inside of students' minds. This dualistic view separates the mind from the world, and can limit our ability to fully comprehend human activity.

This dualistic view has its origins in the work of Descartes, who developed the idea that the mind is fundamentally separate from the body and the world, and that the external world is represented in the mind (Descartes, 1970a, 1970b). This view was adopted by the field of cognitive science in the mid-20th century, with the rise of information-processing models of human thought. For the most part, these models exclude the physical and social contexts from the study of cognition (see Gardner, 1985, for a history of the development of cognitive science).

A growing body of research on situated and embodied cognition has challenged this dualistic view of mind and world and argued that the internal and the external cannot be readily distinguished in human activity (Gibbs, 2005; Lakoff & Johnson, 1999; Varela, Thompson, & Rosch, 1991). Researchers from disciplines ranging from neuroscience to education have argued that thinking, perception, and problem solving are embodied and social activities and that the exclusion of the physical and social contexts from analyses of these activities limits our ability to fully understand them (Goodwin, 2000; Ivarsson, Schoultz, & Säljö, 2002; Lave, 1988; Noble, DiMattia, Nemirovsky, & Barros, 2006; Yamamoto & Kitazawa, 2001).

Studies of students' representations across ages in both science and mathematics illustrate the role of the physical and social contexts in students' construction of meaning (Ehrlén, 2009; Ivarsson et al., 2002; Meira, 2002; Noble, 2007; Noble et al., 2006). For instance, in a study of students' representations of the Earth, Ivarsson and colleagues (2002) found that students interviewed using a globe showed greater understanding of the shape of the Earth and of gravity than students who made their own drawings of the Earth. The authors argued that the availability of the globe as a tool changed the nature of students' reasoning by reminding them of what they knew about the Earth and by "operating as an inference-rich tool" (2002, p. 86) for students. Ivarsson et al. (2002) also reported on an interview study of students' interpretations of a map of the world in which they provided a detailed account of the role of the social interactions between the students and the interviewer in the co-construction of meaning. The authors concluded that the construction of meaning is interactive with the physical and social contexts. Their findings challenge the assumption that students' representations simply reflect some form of internal representation.

REPRESENTATIONS AS TOOLS TO THINK WITH

An alternative to the view that students' representations are reflections of some form of internal representation is the view that representations are among the tools with which we think (Vygotsky, 1978). In this view, which has been elaborated in the literature reviewed in the previous section, scientific content resides neither in the representations nor in the minds of individuals, but in the way the representations are used.

How do students use representations as tools to think with in science? The research on representations reviewed earlier has identified two significant uses for representations in science: *highlighting* and *backgrounding*. These uses of representations have been identified in studies of the practice of science and studies of science and mathematics education. *Highlighting* is a term used by Goodwin (2000) in his work on the use of representations in science. Goodwin described how domain experts use various forms of representation to highlight aspects of the phenomena under study, so that "object[s] of knowledge," that is, things, qualities, or processes, can emerge to be identified, described, and talked about (2000, p. 607). For instance, an archaeologist traced a circular line in the dirt at an archaeological dig site, highlighting for those at the site the difference in coloring between this patch of earth and the surrounding area. This delimited patch of earth could then be talked about as the likely site of a wooden post that had once been at the site, but had since decayed, leaving the patch of darkened earth. The process of creating a representation on the ground that was itself the area under study changed the way that patch of ground was seen by others at the site. In addition, this act of highlighting made present the otherwise absent wooden post that had decayed to create the discoloration. Highlighting by definition makes some aspects of the phenomena under study prominent, and at the same time, it has the potential to bring these aspects into contact with things that are absent.

The use of highlighting has also been documented in research on the representations used by students and teachers to construct meaning in science and mathematics (Noble et al., 2006; Noble, Nemirovsky, DiMattia, & Wright, 2004; Pérez Echeverría & Scheuer, 2009). For instance, Pérez Echeverría and Scheuer (2009) summarized a collection of studies (Andersen, Scheuer, Pérez Echeverría, & Teubal, 2009) in which students' representations allowed them to highlight aspects of a problem or situation and in some cases to highlight the relations among these aspects, such as the relationship between properties of motion and time, facilitating students' reasoning (Robutti, 2009).

Any representation that makes some aspects of phenomena more prominent through highlighting also makes some aspects less prominent, through what I have elsewhere called *backgrounding* (Noble et al., 2006). In a study of high school students using a drawing machine to create a drawing of a circle in a trigonometry class, students created representations that they used as "guide figures" (p. 429) to assist with the complex process of drawing a circle with two independently guided axes controlling the pen's motion (2006, pp. 401–403, provides more details about the device and the mathematics of the activity). Each guide figure highlighted some aspects of the circle, the target of the students' activity, and *backgrounded* others. Each guide figure thus served as a representation that facilitated some aspects of task performance by allowing other aspects to recede into the background.

Backgrounding has also been observed in studies of the representations used by elementary school students when discussing scientific phenomena such as the Earth (Ehrlén, 2009; Ivarsson et al., 2002; Vosniadou, Skopeliti, & Ikospentaki, 2005). Ehrlén (2009) found that students' representations highlighted some aspects of the Earth and backgrounded others, depending upon the purpose of the representation, such as whether the student intended to represent the Earth from a distance versus the Earth from the perspective of an inhabitant. Neither highlighting nor backgrounding is inherently useful or inherently limiting for students; the usefulness of either depends upon the students' goals for the creation and use of a representation. This chapter will explore both backgrounding and highlighting in students' uses of representations as tools for thinking about scientific phenomena.

INVESTIGATING STUDENTS' REPRESENTATIONS OF AIR

This chapter reports on interviews with two students, Catherine and Melanie (both pseudonyms), about how to explain various phenomena involving air using the idea that air is made up of particles. Both students are White, native speakers of English from middle-class households and were students at suburban schools outside of a major U.S. city at the time of the interview. The students were chosen based upon their ages and their availability for interviews and do not constitute a random sample. At the time of their interviews, Catherine was 13 years old and in 8th grade, and Melanie was 9 and in 3rd grade. Neither student considered herself to be particularly successful in science, and neither spontaneously mentioned having previously learned about the particulate nature of matter. However, Melanie and Catherine both knew that air is a gas and Catherine reported having seen a demonstration on air pressure in science class.

The interviews were structured around predetermined tasks (described in more detail in the sections that follow) and questions regarding the behavior of air in different situations, but were open-ended enough to be responsive to the interviewees' goals and questions. The clinical interview technique employed is based upon techniques developed by Piaget and his colleagues (Piaget, 1929/1951), and modified by researchers who developed this tradition (Duckworth, 1970, 1987; Ginsburg, 1997).

The interview protocol and materials were designed to provide opportunities to perceive, discuss, and represent changes in the pressure, volume, and temperature of air[1]. Interviews of this kind have been used extensively in research on students' understandings of scientific phenomena, including gases (Inhelder & Piaget, 1958; Johnson, 1998; Nussbaum, 1985; Piaget, 1927/1965). The episodes and analyses presented in this chapter focus on those instances in which students used representations of the particulate nature of matter in the gaseous phase in the course of their explorations of air.

This chapter presents three short episodes from the interviews: one from Catherine's interview and two from Melanie's interview. The episodes are intended to illustrate some of the possible ways in which students may use representations as they think about the particulate nature of air, and are not intended to be generalizable to a larger group of students.

CATHERINE'S REPRESENTATIONS OF AIR

Catherine chose to be interviewed together with a friend in the same grade, but due to space constraints, this chapter focuses only on Catherine's contributions in the interview. At the time of the interview, Catherine reported finding science class hard.

Episode 1: Drawing Warm and Cold Air

The interviewer brought out several empty 1-liter plastic bottles and initiated an activity in which Catherine stretched a balloon over the top of each bottle and then squeezed the bottle, partially inflating the balloon. Catherine reported having done a related activity in science class, in which a balloon was inflated until it popped. Catherine suggested putting a cap on the bottle, and she found it much more difficult to squeeze the capped bottle than the bottle with a balloon on top. After about 30 minutes of this activity, the interviewer suggested heating and cooling the air inside the bottle by placing the bottle with the balloon on it into a hot or cold water bath (see Figure 7.1).

Catherine then suggested replacing the balloon with a cap before immersing the bottle again in hot and cold water. Catherine found that when she tried to squeeze the capped bottle when it was in hot water, it was slightly harder to squeeze than when it was in cold water.

Scientific Explanation. When one of the bottles is put into hot water, the temperature of the air in the bottle increases, which means that the average speed of the air molecules in the bottle increases. The faster air molecules collide more frequently and with more force with the walls of the bottle and with any covering

Figure 7.1. Bottle with Balloon at Room Temperature (Left), Bottle in a Hot Water Bath (Center), and Bottle in an Ice Water Bath (Right)

on the top of the bottle. When this covering is a balloon, the faster-moving air can push the balloon outward, causing it to inflate slightly, and thus increasing the total volume of the air. When the bottle is capped, the total volume of the air cannot increase significantly, and the result is an increase in the pressure of the air in the bottle, making the bottle harder to squeeze. The air in the cooled bottle experiences the opposite effects.

Right after the experiment with the hot and cold water described above, the interviewer asked:

1. *Interviewer:* Why do the balloons expand when they (the bottles) are in this one (hot water bath) and contract when they are in that one (cold water bath)?

2. *Catherine:* I don't know, I just remember like heat makes things ex-, expand [*two hands move apart*][2] and cold makes things get [pause] smaller [*two hands move closer together*].

   The interviewer then asked Catherine to draw "what the air is doing inside this bottle, when the bottle was warm or when the bottle was cold." Catherine initially responded as follows:

3. *Catherine:* I don't know if just like the air particles (in the heated bottle) would just like, get bigger or something [*right-hand fingers spread*], and

then, uh, the air particles for the colder one would just get much smaller [*right-hand fingertips come together*]?

Catherine made two drawings, re-creations of which are shown in Figure 7.2.

In Catherine's drawings, the hot bottle has 17 large circles in it and the cold bottle has 12 small circles in it. She explained her drawings as follows:

> 4. *Catherine*: Well, for the hot one I just did like, um, the particles, the air particles being like much bigger because they expanded, so like there are more, like, I made it look like there are more of them because when you pressed on something that was hot [*holds right hand as if squeezing bottle*], like, it's much harder to push on it, 'cause, like, it expanded [*two hands move apart*] and stuff, but the cold ones [*points to drawing of cold bottle*], there is less or, they were just much smaller, so you could push on them easier [*right hand makes squeezing motion*].
>
> 5. *Interviewer*: So I have a question. You were saying that you made it look like there were more air particles when it's hot. Is that because you made it look like that because it feels like that, or you made it look like that because it is like that?
>
> 6. *Catherine*: I think just 'cause it like feels like that, 'cause you just get [*right-hand fingertips touch*] like bigger [*right-hand fingers open out*].

Figure 7.2. Re-creation from the Video Record of Catherine's Drawings of Air Molecules in Bottles in Hot (Left) and Cold (Right) Water

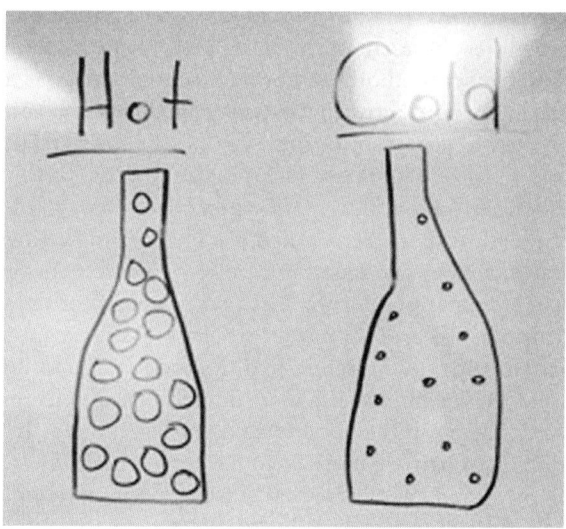

7. *Interviewer*: So when they're bigger it makes it feel like there's more of them in there?

8. *Catherine*: Yeah.

Comments. Catherine expressed uncertainty about what happens to the "air particles" in the hot and cold bottles both before (line 3) and after (lines 4 and 6) she created the drawings re-created in Figure 7.2. She drew more and larger air particles in the hot bottle than in the cold bottle, and said that she "made it look like there are more of them" (line 4), which leaves open the question of whether there actually are more particles in the hot bottle than in the cold bottle. The interviewer asked Catherine directly whether she "made it look like that because it feels like that or . . . because it is like that" (line 5). This statement suggests two alternative intentions for Catherine's drawing, neither of which may have been her original intention. However, Catherine was willing to pick one, answering, "I think just 'cause it feels like that" (line 6). This statement was co-constructed with the interviewer, and is consistent with Catherine's earlier uncertainty about the role and meaning of her drawing as a representation of the air particles.

Catherine's representations were constructed in a social context. For instance, the interviewer's initial question in line 1 focused on the expansion and contraction of the balloons placed on the two bottles. Catherine's drawings and statements reflect a similar focus on expansion and contraction, applied to the air particles themselves. In addition, Catherine used a common textbook convention for the depiction of atoms or molecules as circles. Students frequently use conventional forms of representation for drawings of this type (Dove, Everett, & Preece, 1999), and the medium of drawing, suggested by the interviewer, may be more likely to call upon these conventions than other forms of representation.

Catherine used her representations to *highlight* some of the properties and behavior of air and background others. For instance, her drawing highlighted the difference she felt in pressure between the two capped bottles when one was heated and the other one was cooled, by portraying the particles in the heated bottle as greater in number and size than the particles in the cooled bottle. This instance of highlighting made connections between the increased difficulty of squeezing one bottle, which can be directly perceived, and the changes in the invisible particles of air, which cannot, thereby constructing a cause for the felt change in the bottle.

Catherine's representations *backgrounded* other aspects of the behavior of air, such as the intrinsic motions of air particles. Catherine depicted the particles as stationary, rather than including lines or arrows indicating motion. This type of representation, and the medium of drawing, limited her options for depicting the differences between the air in the hot and cold bottles, and may have contributed to the backgrounding of particle motions in her representations.

Catherine's uncertainty about her verbal and drawn representations suggests that she was using them to propose some ideas about why the hot and cold bottles felt different to her. She used her drawing, utterances, and gestures to imagine and explore size and number of particles as possible causes of this difference.

MELANIE'S REPRESENTATIONS OF AIR

At the time of the interview, Melanie considered herself to be a good student overall, but not particularly successful in science. The two episodes from Melanie's interview that follow include four different representations of air that Melanie used as she explained the behavior of air in various situations. These four representations are: a verbal analogy between air molecules in a bottle and people at a party, a drawing of air molecules in a bottle, a verbal analogy between air molecules and solid objects, and a verbal analogy between air molecules in a syringe and LEGO® bricks in a cup.

Episode 1: People at a Party

Melanie's interview also began with the task of squeezing a plastic bottle with a balloon stretched over the top of it. As Melanie began to engage with this task, she suggested an alternative task. She wished to get air from the balloon to go into the bottle by placing a fully inflated balloon on top of the bottle. Melanie blew up the balloon and placed it over the top of the bottle, but the balloon remained fully inflated (see Figure 7.3), rather than sending its air into the bottle as she had anticipated.

Melanie explained:

1. *Melanie*: Um, it doesn't work. I thought that if you put the air[3] in the balloon, it's attached to the bottle and would go into the bottle, but the bottle is already full of air, and since there's room in the balloon that you can put the air, it's just, staying where it is. . . .
2. *Interviewer*: So, so say I'm a little air molecule [*points at balloon*] and I'm in here [*points at balloon*] and then I come down here [*points at bottle*], just I'm wandering around [*circling motion with finger*] and I come down here [*points at bottle*] and I want to go into this bottle [*points at bottle*], what, what's keeping me from going in [*points at bottle*]?
3. *Melanie* Well, I think all the, ah ( )[4] the other little [*fingers wiggle near balloon*], whatever they're called?
4. *Interviewer*: Air molecules.
5. *Melanie*: Air molecules are having a big party down here [*hands on bottle*] and they've shhh [*two hands make "shut-off" gesture, moving together and apart, palms down, horizontal*] the room is already full [*shut-off gesture*] and you can't go in! . . .

6. *Melanie*: [I]f you're at a party where it's crowded, though, and it's hard to
   move around, but you <u>can</u> do it, I think that's what it's like, but if y-, if
   there's, there's only so much space [*hands move in arcs out to sides*], and if
   you have, it's already pretty full [*shut-off gesture*] like if there's lots of people
   in there and it's hard to move, but you <u>can</u> do it, [technical interruption].
   Okay, so, it's like, there <u>is</u> room to move around, but if you put in another
   person, there wouldn't be as much room to move around in.

After a few minutes more of discussion, the interviewer asked Melanie to draw
what the air molecules look like in the bottle (see Figure 7.4).

Comments. To explain why the air from the balloon stayed in the balloon
instead of moving down into the bottle, Melanie described the bottle as being
"*full*" of air, so that no additional air would fit in the bottle.

In line 2, the interviewer introduced the term "*air molecule*" and invited Melanie to use an analogy between a person and an air molecule to explain why the
air from the balloon could not enter the bottle. Melanie elaborated upon the interviewer's analogy by describing the air molecules in the bottle as people at an
overcrowded party who were keeping the air molecules/people from the balloon
out. The idea of the party may have originated in Melanie's suggestion earlier in
the interview that a party store would be a good place to buy better balloons. The
construction of this analogy was distributed between the interviewer and Melanie,
and was influenced by the details of the physical and social contexts.

**Figure 7.3. An Inflated
Balloon Placed on Top
of an "Empty" Plastic
Bottle**

**Figure 7.4. Melanie's Drawing of Some
of the Air Molecules in the Bottle**

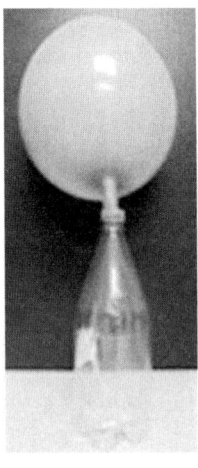

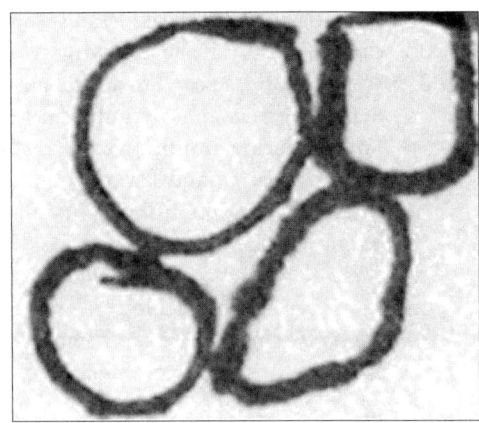

The analogy between people at a party and air molecules in the bottle *highlighted* the ability of air to fill up the space in the bottle and to keep the air from the balloon out, the unexpected behavior she observed in this episode. The people-at-a-party analogy connected the macroscopic behavior of the air in the bottle and balloon with both the invisible air and visible people at a party. The analogy thus allowed Melanie to explore a possible mechanism by which the air molecules in the bottle can take up space and keep other molecules out: by being closely packed like people at an overcrowded party and physically taking up nearly all the available space.

In addition to highlighting the ability of the air to fill space, the people-at-a-party analogy also *backgrounded* other properties of air that Melanie was likely to be less familiar with. For instance, air molecules at standard temperature and pressure (STP, 20°C and 1 atmosphere of pressure) take up only about one-thousandth of any volume they are in, meaning that 999/1000ths of any space filled with air at STP is empty space. In addition, air molecules at STP move at very high average speeds and frequently collide with each other and the container walls. In the people-at-a-party analogy and the drawing Melanie made of air molecules (Figure 7.4), there is very little empty space or motion. As in Episode 1 from Catherine's interview, the medium of drawing and Melanie's choice of circles to represent air molecules contributed to the backgrounding of the motion of the molecules.

In the next episode, we will see how Melanie engaged with another representation of air.

Episode 2: LEGOs® in a Cup

In the 48 minutes between the end of Episode 1 and the start of Episode 2, Melanie and the interviewer continued to discuss and elaborate upon the people-at-a-party analogy and several other ways of describing air in other contexts.

The interviewer brought out a plastic syringe (consisting of a plunger and a plastic cylinder) and a rubber stopper and asked Melanie to try pressing down on the plunger to force the air through the end of the cylinder. The interviewer then asked Melanie to try pressing on the plunger with the rubber stopper blocking the end of the cylinder (see Figure 7.5).

Melanie was only able to push the plunger down part of the distance to the end of the syringe (see Figure 7.5 for an example), and when the interviewer asked her why, Melanie answered as follows:

7. *Melanie*: [*replaces plunger at the top of the syringe on the stopper*] That, the, um, you can only squeeze it so much because the air molecules, even if they <u>can</u> be pushed together, like most things can, like I can push these two things together [*pushes plastic bottle and weighted plastic*

tape dispenser together], but only so far. (*Interviewer*: Okay.) Say you are pushing down on the molecules [*presses down on plunger*], and so say you are pushing down on a mixture of these objects and [*picks up bottle, points to the tape dispenser, a cup*] cups, and all that stuff, and, um, you, push down [*presses down on plunger again*], they can only go so far because, I mean [sighs]. You know when you go to that, there's that LEGO® store at the mall?

8. *Interviewer*: I've never been there, but tell me about it.

9. *Melanie*: Okay. In the mall, and there's this, and in the back [*removes plunger from syringe and places plunger and syringe on table*] there's these little like shelves [*hand gestures to side*] of little, um, LEGOs® that you can choose, and you get these cups [*hand makes cup shape*], and [pause] if you pack it <u>real</u> nice, like put a layer down [*flat hand makes spreading motion across table*], you're gonna get more <u>than</u> if you just throw them in pell-mell [*mimes throwing objects into container*]. <u>So</u>, it's sort of like this [*sets up syringe and stopper again*] and then, you decide, I really want to get as <u>many</u> (LEGOs®) as I can. And these blocks are like cement, like the inside of this [*picks up weighted tape dispenser*] I think. So you can<u>not</u> like break it at all. (*Interviewer*: Okay, they're really solid.) <u>Really</u> solid. And you push down [*pushes on plunger a little*], they're gonna sift [*spread hand shakes a little from side to side*]. That's sort of I think what happens to the air molecules.

10. *Interviewer*: So do you think it's kind of sifting [*spread hand shakes side to side*]?

11. *Melanie*: Until it gets [*hand grasps syringe and slides down to stopper*] so it (plunger) can't move at all, and that's when it stops [*presses on plunger*].

Comments. The activity of compressing the air in the syringe was designed to demonstrate both the compressibility of the air and the limits on its compressibility. To explain what she observed, Melanie first made an analogy between molecules of air and some solid objects she found nearby in the interview room—a plastic bottle, a weighted plastic tape dispenser, and a paper cup—to suggest how the air molecules might be pushed closer together, but only so much. Apparently unsatisfied with this analogy, she created a new analogy between air molecules in a syringe and LEGO® bricks in a cup. According to Melanie, when filling a cup with LEGO® bricks, there are strategies one can use to get more bricks into the cup. For instance, packing the bricks in "real nice" and putting a "layer down" allows you to get more bricks than throwing them in "pell-mell" (line 9). This phenomenon is illustrated by the picture in Figure 7.6 of two cups assembled by the author after the interview. Each cup contains an identical collection of LEGO® bricks, but the bricks in the cup on the left were loosely thrown in, while those in the cup on the right were packed more carefully.

Figure 7.5. The Syringe and Rubber Stopper with the Plunger Depressed as Far as the Author Can Push It

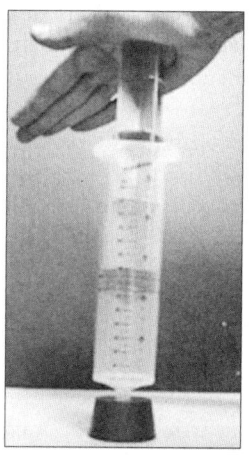

Figure 7.6. Two Cups of Identical LEGO® Bricks, One in Which They Are Thrown in Loosely (Left), and One in Which They Are Packed Carefully (Right)

The LEGOs®-in-a-cup analogy allowed Melanie to highlight the compressibility of air by explaining that air molecules can "sift" (line 9), the way that LEGO® bricks do as they become more compact in a cup when they are moved around. Unlike the people-at-a-party analogy, in which no more people (or molecules) could fit in, the LEGO® bricks analogy allows for the compressibility of air, because the bricks can fit into a smaller space if they more optimally arranged. This can be achieved either by placing them carefully ("you pack it real nice," line 9), or by shaking the container in ways that allow the pieces to settle ("they're gonna sift," line 9). While highlighting the compressibility of air, Melanie also made a connection between the invisible air and visible, graspable LEGO® bricks. This connection between the visible and the invisible allowed Melanie to explore a possible mechanism for the compression of air: the movement of molecules into more optimal arrangements.

HOW THE STUDENTS USED REPRESENTATIONS OF AIR

Melanie used three different analogies to represent air molecules: people-at-a-party, solid objects, and LEGOs®-in-a-cup. The use of multiple analogies is common among scientists (Shelley, 2003). For instance, the analogy between light and a wave is an effective way to explain the behavior of light in some circumstances, and in other circumstances light behaves more like a particle and the wave analogy cannot be used. Other examples include the analogy between heat and a fluid substance and the analogy between atoms in a gas and billiard balls, both of which are useful in some cases and not in others (Frigg & Hartmann, 2009; Marion & Hornyak, 1982). None of these analogies reflects a comprehensive model of the phenomenon it describes. Instead, each one highlights particular aspects of the phenomenon and provides a means for thinking about the mechanism whereby it occurs, while backgrounding other aspects of the phenomenon to allow focus on the highlighted aspects.

Catherine and Melanie used representations both to highlight particular aspects of the behavior of air, such as its ability to expand and to fill space, and to make connections between visible phenomena they observed and invisible particles of air. In addition, each of their representations of air backgrounded other aspects of air, such as the motion of air particles. Elementary and middle school students' representations of gases often do not include the motion of gas molecules, and other researchers have taken this as an indication that students' internal representations of gases do not include the motions of gas particles (Nussbaum, 1985; Nussbaum & Novick, 1982). However, the perspective developed in this chapter is that the backgrounding of gas particle motion, rather than diagnosing some deficit in students' internal representations, instead reflects the fact that other aspects of the behavior of air were more salient to the students as they performed these tasks, in these physical and social contexts. An important question to explore for future work is how varying the physical and social contexts for students' exploration of scientific phenomena may influence those aspects of the phenomena that students highlight in their own representations (see Chapter 9).

Both Catherine and Melanie used representations to explore possible underlying causes for the behaviors they observed, even when they were uncertain about these causes. They used representations to imagine how the air molecules that they could not see or feel caused phenomena that they could see and feel, such as the air becoming difficult to compress further in the plastic cylinder. The students' representations served as spaces for exploring ideas, by allowing them to imagine the air as visible circles in a drawing or as visible people at a party. The use of representations as tools for imagining is a practice that has been described in both the practice of science (Ochs, Gonzales, & Jacoby, 1996; Ochs, Jacoby, & Gonzales, 1994; Wolpert & Richards, 1997) and in science education (Ehrlén, 2009; Noble, 2007; Rosebery, Ogonowski, DiSchino, & Warren, 2010; Warren, Ballenger, Ogonowski, Rosebery, & Hudicourt-Barnes, 2001). Consistent with the findings of the studies mentioned above, Catherine and Melanie used their representations not only to highlight and background the observable, but also to begin the imaginative process of exploring the unseen causes of observable phenomena.

The interview context offered Catherine and Melanie space to explore their own abilities to imaginatively construct representations of air, without having these representations evaluated for their fidelity to a standard scientific representation. Both students engaged with the task of representing air, and encountered some of the strengths and limitations of any representation of a physical phenomenon. Similar experiences of creatively constructing representations, if incorporated into school science, could enhance students' understandings of the nature and purpose of scientific representations of physical phenomena.

ACKNOWLEDGMENTS

This research has been supported by the Fulcrum Institute project (NSF-MSP 01256) and The Education Research Collaborative at TERC. Opinions expressed herein are those of the author and not necessarily those of the National Science Foundation (NSF) or The Education Research Collaborative. The author thanks Catherine, Melanie, and all the other students, teachers, and physicists who participated in this study, for all that they have taught her. The author also thanks David Carraher, Paul Wagoner, and Chunhua Liu for their collaboration on this research; Ann Rosebery for many insightful conversations regarding drafts of this chapter; and the editors of this volume, one anonymous reviewer, and the Teachers College Press editors for their thoughtful and constructive comments and suggestions on earlier versions of this chapter.

NOTES

1. The interviews reported on in this chapter came from a larger study of students' ideas about pressure, volume, and temperature that included a total of eight students of varying ages.

2. The bracketed text describes a gesture or other action.

3. Underline indicates stress. Double underline indicates emphatic stress.

4. Words that could not be interpreted from the recording.

REFERENCES

Acher, A., & Arca, M. (2009). Children's representations in modeling scientific knowledge construction. In C. Andersen, N. Scheuer, M. P. Pérez Echeverría, & E. V. Teubal (Eds.), *Representational systems and practices as learning tools* (pp. 109–131). Rotterdam, Netherlands: Sense Publishers.

Adadan, E., Irving, K. E., & Trundle, K. C. (2009). Impacts of multi-representational instruction on high school students' conceptual understandings of the particulate nature of matter. *International Journal of Science Education, 31*(13), 1743–1775.

Andersen, C., Scheuer, N., Pérez Echeverría, M. P., & Teubal, E. V. (2009). *Representational systems and practices as learning tools.* Rotterdam, Netherlands: Sense Publishers.

Crowder, E. M. (1996). Gestures at work in sense-making science talk. *The Journal of the Learning Sciences, 5*, 173–208.

Descartes, R. (1970a). Discourse on the method of rightly conducting the reason and seeking for truth in the sciences. In E. S. Haldane & G. R. T. Ross (Eds.), *The philosophical works of Descartes* (Vol. 1, pp. 81–143). Cambridge, UK: Cambridge University Press.

Descartes, R. (1970b). Rules for the direction of the mind. In E. S. Haldane & G. R. T. Ross (Eds.), *The philosophical works of Descartes* (Vol. 1, pp. 1–77). Cambridge, UK: Cambridge University Press.

Dove, J. E., Everett, L. A., & Preece, P. F. W. (1999). Exploring a hydrological concept through children's drawings. *International Journal of Science Education, 21*(5), 485–497.

Duckworth, E. (1970). Piaget rediscovered. *The ESS Reader* (pp. 135–139). Newton, MA: Educational Development Center.

Duckworth, E. (1987). *"The having of wonderful ideas" and other essays on teaching and learning.* New York: Teachers College Press.

Ehrlén, K. (2009). Drawings as representations of children's conceptions. *International Journal of Science Education, 31*(1), 41–57.

Frigg, R., & Hartmann, S. (2009). Models in science. *The Stanford encyclopedia of philosophy.* Retrieved March 10, 2011, from http://plato.stanford.edu/archives/sum2009/entries/models-science/

Gardner, H. (1985). *The mind's new science: A history of the cognitive revolution.* New York: Basic Books.

Gentner, D., & Gentner, D. R. (1983). Flowing waters or teeming crowds: Mental models of electricity. In D. Gentner & A. L. Stevens (Eds.), *Mental models* (pp. 99–129). Hillsdale, NJ: Lawrence Erlbaum Associates.

Gibbs, R. W. (2005). *Embodiment and cognitive science.* Cambridge, UK: Cambridge University Press.

Ginsburg, H. P. (1997). *Entering the child's mind: The clinical interview in psychological research and practice.* New York: Cambridge University Press.

Goodwin, C. (2000). Action and embodiment within situated human interaction. *Journal of Pragmatics, 32*, 1489–1522.

Inhelder, B., & Piaget, J. (1958). *The growth of logical thinking from childhood to adolescence: An essay on the construction of formal operational structures* (A. Parsons & S. Milgram, Trans.). New York: Basic Books.

Ivarsson, J., Schoultz, J., & Säljö, R. (2002). Map reading versus mind reading. In M. Limón & L. Mason (Eds.), *Reconsidering conceptual change: Issues in theory and practice* (pp. 77–99). Dordrecht, Netherlands: Kluwer Academic.

Johnson, P. (1998). Progression in children's understanding of a "basic" particle theory: A longitudinal study. *International Journal of Science Education, 20*(4), 393–412.

Lakoff, G., & Johnson, M. (1999). *Philosophy in the flesh: The embodied mind and its challenge to Western thought.* New York: Basic Books.

Lave, J. (1988). *Cognition in practice: Mind, mathematics, and culture in everyday life.* Cambridge, UK: Cambridge University Press.

Marion, J. B., & Hornyak, W. F. (1982). *Physics for science and engineering: Parts 1 & 2 combined.* Philadelphia, PA: Saunders College.

Meira, L. (2002). Mathematical representations as systems of notations-in-use. In K. Gravemeijer, R. Lehrer, B. van Oers, & L. Verschaffel (Eds.), *Symbolizing, modeling and tool use in mathematics education* (pp. 87–103). Dordrecht, Netherlands: Kluwer Academic.

Noble, T. (2007). *Body motion and physics: How elementary school students use gesture and action to make sense of the physical world.* Unpublished doctoral dissertation, Tufts University, Medford, MA.

Noble, T., DiMattia, C., Nemirovsky, R., & Barros, A. (2006). Making a circle: Tool use and the spaces where we live. *Cognition and Instruction, 24*(4), 387–437.

Noble, T., Nemirovsky, R., DiMattia, C., & Wright, T. (2004). Learning to see: Making sense of the mathematics of change in middle school. *International Journal of Computers for Mathematical Learning, 9*(2), 109–167.

Nussbaum, J. (1985). The particulate nature of matter in the gaseous phase. In R. Driver, E. Guesne, & A. Tiberghien (Eds.), *Children's ideas in science* (pp. 124–144). Milton Keynes, UK: Open University.

Nussbaum, J., & Novick, S. (1982). Alternative frameworks, conceptual conflict and accommodation: Toward a principled teaching strategy. *Instructional Science, 11*, 183–200.

Ochs, E., Gonzales, P., & Jacoby, S. (1996). "When I come down, I'm in the domain state": Grammar and graphic representation in the interpretive activity of physicists. In E. Ochs, E. A. Schegloff, & S. Thompson (Eds.), *Interaction and grammar* (pp. 328–369). Cambridge, UK: Cambridge University Press.

Ochs, E., Jacoby, S., & Gonzales, P. (1994). Interpretive journeys: How physicists talk and travel through graphic space. *Configurations, 2*(1), 151–171.

Pérez Echeverría, M. P., & Scheuer, N. (2009). External representations as learning tools: An introduction. In C. Andersen, N. Scheuer, M. P. Pérez Echeverría, & E. V. Teubal (Eds.), *Representational systems and practices as learning tools* (pp. 1–17). Rotterdam, Netherlands: Sense Publishers.

Piaget, J. (1965). [1927]. *The child's conception of physical causality* (M. Gabain, Trans.). Totowa, NJ: Littlefield, Adams.

Piaget, J. (1951). [1929]. *The child's conception of the world* (J. Tomlinson & A. Tomlinson, Trans.). Savage, MD: Littlefield Adams.

Pozo, J. I., & Gómez Crespo, M. A. (2005). The embodied nature of implicit theories: The consistency of ideas about the nature of matter. *Cognition and Instruction, 23*(3), 351–387.

Robutti, O. (2009). Space-time representations in young children. In C. Andersen, N. Scheuer, M. P. Pérez Echeverría, & E. V. Teubal (Eds.), *Representational systems and practices as learning tools* (pp. 59–76). Rotterdam, Netherlands: Sense Publishers.

Rosebery, A., Ogonowski, M., DiSchino, M., & Warren, B. (2010). "The coat traps all your body heat": Heterogeneity as fundamental to learning. *Journal of the Learning Sciences, 19*(3), 322–357.

Shelley, C. (2003). *Multiple analogies in science and philosophy.* Amsterdam, Netherlands: John Benjamins.

Varela, F. J., Thompson, E., & Rosch, E. (1991). *The embodied mind: Cognitive science and human experience.* Cambridge, MA: MIT Press.

Vosniadou, S., & Brewer, W. F. (1992). Mental models of the earth: A study of conceptual change in childhood. *Cognitive Psychology, 24,* 535–585.

Vosniadou, S., Skopeliti, I., & Ikospentaki, K. (2005). Reconsidering the role of artifacts in reasoning: Children's understanding of the globe as a model of the earth. *Learning and Instruction, 15*(4), 333–351.

Vygotsky, L. S. (1978). *Mind in society: The development of higher psychological processes.* Cambridge, MA: Harvard University Press.

Warren, B., Ballenger, C., Ogonowski, M., Rosebery, A., & Hudicourt-Barnes, J. (2001). Rethinking diversity in learning science: The logic of everyday sense-making. *Journal of Research in Science Teaching, 38,* 1–24.

Wolpert, L., & Richards, A. (1997). *Passionate minds: The inner world of scientists.* Oxford, UK: Oxford University Press.

Yamamoto, S., & Kitazawa, S. (2001). Sensation at the tips of invisible tools. *Nature Neuroscience,* 979–980.

CHAPTER 8

Diagram-Use and the Emergence of Mathematical Objects

Ricardo Nemirovsky
and Michael Smith

MOST MATHEMATICAL INSCRIPTIONS ARE PEIRCEAN DIAGRAMS

In this chapter we analyze utterances by a mathematician as he explains a paper that he had published in a major mathematical journal. Over the course of the interview he describes certain symmetries in the algebraic structures that correspond to the combination of spins in quantum mechanics; it turns out that these symmetries are geometrically related to those of a tetrahedron. Our intent in this examination is to explore the ways in which the inscriptions, gestures, and speech of the mathematician created or revealed mathematical entities for him and the interviewer. As support for our analysis, we draw on selected core ideas about diagrams and mathematical practices elaborated by Charles Peirce. Rather than trying to outline an overarching Peircean perspective, which encompasses a complex and sprawling array of outlooks and expressions that resist executive summaries or compact categorizations, we identify just a few of Peirce's thoughts about representations and mathematics that provide, we think, an insightful demarcation of the work with mathematical signs within the general landscape of signs of all types. In pursuing this goal, the writings of Stjernfelt (2000, 2007), Campos (2010), and Lehrer, Strom, and Confrey (2002) greatly assisted us.

Peirce (1991) proposed classifying signs into three kinds:

> Every sign is determined by its object, either first, by partaking in the characters of the object, when I call the sign an *Icon*; secondly, by being really and in its individual existence connected with the individual object, when I call the sign an *Index*; thirdly, by more or less approximate certainty that it will be interpreted as denoting the

object, in consequence of a habit, when I call the sign a *Symbol*. (p. 251; emphasis in the original)

To state this somewhat differently: *Symbols*, such as the words *triangle* or *perimeter*, refer to their referents by means of conventions, habits, and cultural practices. *Indices* relate factually or empirically to the particular qualities of the denoted object; for instance, both phrases "I see a man with a rolling gait. This is a probable indication that he is a sailor" and "a low barometer with a moist air is an index of rain" (Peirce, 1931–1935, notes 285–286) exemplify the use of indices. Note that a low barometer with a moist air in a closed humid room does not indicate rain, which suggests the crucial point that indices achieve their deictic function by incorporating the context in which they are used (e.g., using a barometer in open air or in a closed room alters its indexicality). Letters that index components of a geometric figure offer another example of this contextual dependency, so that direct or indirect access to the drawing in which the letter is being used is necessary to know what it refers to (Netz, 1999). *Icons*, which include maps and photographs, are signs that relate to their object by similarity.

Stjernfelt (2000) elaborated on how Peirce's conception of what counts as "similar" was a key contribution in semiotics because it addressed a well-known criticism: given that "any phenomenon can be said to be like any other phenomenon in some respect"(Stjernfelt, 2007, p. 28), the concept of similarity as a generic likeness turns out to be empty. According to Peirce, instead, similarities between an icon and its referent manifest in the opportunities to experiment and transform the icon in order to reveal new or nontrivial qualities of the referent. Stjernfelt (2000) describes iconic similarity in these terms: "an icon is characterized by containing implicit information [about its referent] which in order to appear must be made explicit by some more or less complicated procedure accompanied by observation" (p. 359). For instance, that we can use a map to calculate the distance between two locations reflects a relationship of similarity between the map and the surveyed territory. While there is nothing relevant to learn about individuals by, say, measuring the distance between the characters within each of their names, there are countless aspects that can be learned about a piece of music by studying its score; these latter possibilities show that a score refers to a piece of music through a common structure. In actual use, this type of similarity between an icon and its referent often turns into a complete substitution:

> Icons are so completely substituted for their objects as hardly to be distinguished from them. . . . So in contemplating a painting, there is a moment when we lose the consciousness that it is not the thing, the distinction of the real and the copy disappears, and it is for the moment a pure dream, not any particular existence, and yet not general. At that moment we are contemplating an icon. (Peirce, 1885, p. 181)

Whether the referent is real, fictional, hypothetical, or far-fetched is immaterial to the iconic use of a sign. For instance, just as we can examine a sketch of an actual house in order to notice details we might have overlooked, we can do the same with a sketch of a fictional house despite the referent of the icon being a fantasy. Thus, icons offer a physical form that we can substitute for unperceivable objects offering something material with which to interact.

Consistent with his apparent inclination to propose triads, Peirce catalogued icons into three types: Images, Diagrams, and Metaphors. The similarity between Images and their objects is available by direct perceptual inspection, such as in a portrait of a person. Diagrams, such as the organizational chart of an institution, "resemble their objects not [necessarily] at all in looks; it is only in respect to the relations of their parts that their likeness consists" (Peirce, 1931–1935, note 282). Although this is admittedly counterintuitive (at least for an English speaker), Peirce described algebraic equations as instances of Diagrams: "every algebraical equation is an icon, in so far as it exhibits, by means of the algebraic signs (which are not themselves icons [such as the signs + or =]), the relationships of the quantities concerned" (Peirce, 1931–1935, note 282; brackets added). The iconicity of an algebraic equation and its referent, which defines its iconic character, opens up possibilities for manipulating the equation to reveal nontrivial properties of the quantities or functions in question: "the reasoning of mathematicians will be found to turn chiefly upon the use of likeness, which are the very hinges of the gates of their science. The utility of likeness to mathematicians consists in their suggesting in a very precise way, new aspects of supposed state of things" (Peirce, 1931–1935, note 281). For instance, altering the linear equation $y = 2x + 8$ so that it reads as $y = 2(x + 4)$ can make the location of the corresponding line's x-intercept more directly visible to an algebra student.

Notice that in this last example it is necessary to say for whom the Diagram manipulations prove to be insightful. Someone who is totally ignorant of algebraic expressions would have a hard time recognizing the equations $y = 2x + 8$ and $y = 2(x + 4)$ as describing the same relationship between x and y, let alone be able to see anything new in the second that might be unclear in the first. At the other extreme, someone who is highly proficient with algebra, such as a mathematician, would probably see the potential for altering one equation to become the other so readily that actually performing that alteration would probably not yield any new insights for him or her. This is a specific example of the general use of Diagrams: Both the manipulations that appear to the individual as viable as well as the ability for those manipulations to offer insight depend critically on the prior experience that the user has had with Diagrams of the sort in question.

Stjernfelt (2000) shows that Peirce distinguished between "empirical" and "pure" Diagrams in ways that resemble the Kantian distinction between a-posteriori and a-priori relations, respectively. A Diagram indicating that the population of a certain country is twice that of the population of another country is likely to be used to indicate an empirical relation (i.e., observational, based on data collection), whereas a Diagram illustrating that the diameter of a circle is twice the length of its radius tends to express a pure one (i.e., postulated, definitional). This distinction is particularly relevant for our work because mathematics, at least in the domains characterized under "pure mathematics," is coextensive with the use of pure Diagrams (Stjernfelt, 2000).

The icons that Peirce grouped under "Metaphors" invoke a third, mediating object through which the icon resembles its referent. At least in the contexts of science and mathematics, Metaphors often rely on Diagrammatic features. For instance, shortly after Rutherford's discovery of the atomic nucleus, atoms were described as small "planetary systems." The power of this Metaphor centered on the possible use of Diagrams depicting gravitational attraction, which, like electric forces, vary with the inverse of the distance squared, to account for electrons orbiting the nucleus. Another example is the Metaphor of particle-as-wave, which is reflected by Schrödinger's equation having the form of a wave equation. While the Metaphor of atom-as-planetary-system has been abandoned, the one of particle-as-wave has become a cornerstone of Diagrammatic work in quantum mechanics.

THE MEANING OF DIAGRAMS IS THEIR USE

In this section we outline an approach that is relevant to the present chapter by extending Peirce's conception of Diagrams with ideas stemming from our prior work (e.g., Nemirovsky, 1994). Our focus is Diagram-use, by which we mean the actual use of Diagrams by one or more people in a specific place and time who are engaged in real or virtual interactions with others (Nemirovsky, 1994). This means that we treat Diagrams not merely as inscriptions recorded on a surface but as inseparable from enacted or implicit gesture, talk, drawing, body movement, and so forth, as well as feelings, expectations, and personal identities. This emphasis extends Peirce's: It amounts to viewing icons as being constituted by enacted or implicit gestures, gaze, bodily poise, or talk, even when nothing relevant is depicted on surfaces surrounding the participants, in which case the icon itself is imaginary.

Our conception of Diagram-use contravenes the opinion based on the common observation that Diagrams recorded on a surface seem to stand "on their own" such as, say, when we open a book with Diagrams drawn on its pages: We might browse the book and see inscribed Diagrams without anyone talking, pointing, or gesturing about them. However, if we were to ask what these Diagrams mean to a reader at a certain time, we would have to envision the uses projected by the reader and perhaps suggested by the text. Perceiving the manipulability of, or interpreting, a Diagram amounts to achieving bodily readiness, even if covert, for appropriate kinds of talk, gesture, gaze, and action (Noble, Nemirovsky, DiMattia, & Wright, 2004). What counts as "appropriate" in this sense emerges from a history of engagements with cultural practices in which such interpretations of the Diagram are customary. For this reason, the idea that a Diagram inscribed on a surface has meaning on its own in a way that is detached from users' actual or potential perceptuo-motor activity (gestures, talk, drawings, body positioning, and so on) is an illusion derived from the fact that in many circumstances the latter activity is tacit.

In the episode we examine, the mathematician elaborates on the addition of angular momentum in quantum mechanics. In his Diagrams, the edges look like arrows because they are directional, with each corresponding to the angular momentum of a particle denoted by the letter J. Each node corresponds to a combination of two angular momentums, leading to a third resulting angular momentum. Figure 8.1 is an example where J1 + J2 = J4 and J4 + J3 = J5.

HOW AND FROM WHOM WE COLLECTED OUR DATA

The subject for our study was a mathematician whom we will refer to as "Joseph." We asked Joseph to pick a published paper of his that he considered significant and would be willing to discuss with us. Once he had done this, we (the authors) reviewed his paper and formulated questions to guide the subsequent discussion with Joseph. We then conducted a 1-hour unstructured interview (Bernard, 1988) with Joseph, which we recorded with two digital cameras positioned at opposite sides of the room. By and large, most of the interview consisted of Joseph explaining the paper and its surrounding ideas to us rather than being a question-and-answer session.

Figure 8.1. Combination of Angular Momentum, J1 + J2 + J3 → J5

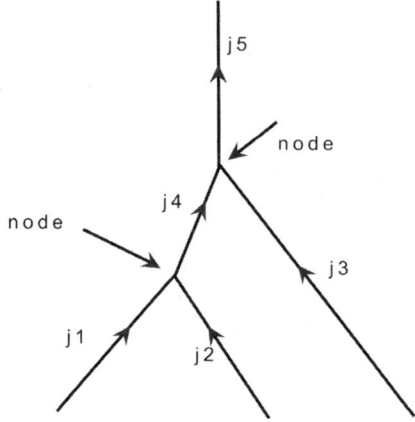

We reviewed the videos of the interview several times and chose segments upon which to perform what Erickson (2004) refers to as multimodal microanalysis. A multimodal microanalysis attends to multiple streams of activity such as tone of voice, facial expression, gesture, gaze, and so forth. Our selections were based on which segments appeared to offer the greatest potential insight into a number of key phenomena, including Diagram-use. The product of microanalysis for each segment is a document with annotated screenshots indicating movement, timestamps corresponding to the screenshots, and a transcript that is time-coordinated with the screenshots such that each portion of transcript is tied to the time interval between screenshots. This allowed us to review each selection in great detail. Finally, we coordinated the insights produced by these microanalysis documents with the mathematical content to which Joseph referred in order to produce a narrative that we believe offers significant insight into the nature of Diagram-use.

The paper Joseph chose was one he had published 10 years prior to our interview with him and focused on an area of overlap between topology and quantum physics. In classical physics, there is a straightforward rule based on vector addition for combining the rate at which two objects are spinning in order to determine their net spin. However, the situation is more complicated in quantum physics, requiring the use of what is known as the 6j symbol in order to compare different spin combinations. While the 6j symbol is a computational tool, Joseph's paper elaborated upon an interpretation of the 6j symbol that allowed it to be understood in terms of a Euclidean three-dimensional tetrahedron, which in turn revealed geometric symmetries in the computations that had previously gone unnoticed in his field.

EXAMPLES OF DIAGRAM-USE FROM A MATHEMATICIAN

In what follows, we will describe a selection from the interview with Joseph in which he constructed a tetrahedron by comparing two graphs showing how to combine the spins of three particles. We will outline this by illustrating four utterances and offering our commentary on them. Underlined segments of the transcribed utterance indicate that they occurred at the same time as the gesture marked in the accompanying image. Utterance 1 relies on some context provided by Figures 8.2, 8.3, and 8.4.

Earlier in the interview, Joseph had drawn the tree Diagram in Figure 8.2 in order to show how to combine three particles' spins in pairs: First, j_1 and j_2 are added together to make j_4, and then j_4 and j_3 are added together to produce j_5. One of us (Michael) then drew the Diagram in Figure 8.3 and asked whether it ever happens that the three spins get added all at once. This prompted Joseph to erase the juncture in Michael's Diagram and to cover it with a surrounding "black box"

Figure 8.2. Initial Addition of j1, j2, and j3

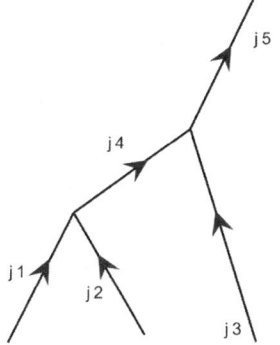

Figure 8.3. Michael's Question

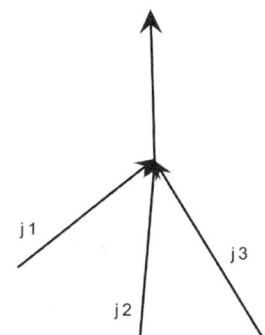

Figure 8.4. The Black Box

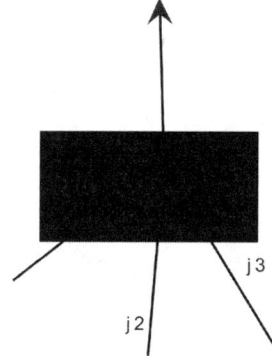

(see Figure 8.4). This black box put out of sight the particulars about how the three vectors append to each other, freeing up the consideration of alternatives. Joseph clarified that all one cares about is that the three spins combine in some manner to produce a new conglomerate spin. This led to Utterance 1 (see Figure 8.5).

In order to make the case for binary Diagrams, in which vectors join by pairs, Joseph introduced an intermediary component between vector Diagrams and interacting particles: the "simple beings" who are calculators of angular momentum addition. "We" are "sort of simple beings" [U1-1] who have "learned how to combine two guys to get one" [U1-2, U1-3]. The simplicity consists of wanting to reduce a complex case into a composition of simpler and known cases. We propose that this reference to mathematicians as "simple beings" makes use of a Peircean Metaphor. Note that what we here propose as a Metaphor— i.e., as a "third" mediating element—is that vector additions can be composed in

Figure 8.5. Utterance 1

[U1-1] But since we're, we're, we're, we--'re sort of simple beings,

[U1-3] two guys to get one

[U1-2] We've <u>learned</u> how to combine

[U1-4] So we just say okay if I wanna combine three guys; let me

[U1-6] rather than being over<u>ambitious</u>.

[U1-5] just do it in <u>two stages</u>

terms of pair-additions, which are preferable because for "us" they are "familiar" (see Figure 8.6).

As Joseph came to refer to himself as a member of a certain community (i.e., from "we" to "let me" [U1-4]), he turned his gaze away from the blackboard, looked at Michael, smiled, and gestured a two-stage solution by circling his left hand twice [U1-5]. His circling gesture made explicit the generality he sees in the argument: The addition of any set of three angular momentum vectors can be solved in two stages. Such foregrounding of generalizability is possibly a common motivation to invoke explicit Metaphors: They allow us to group an infinite number of cases under common and nontrivial qualities. Joseph then opposed the two-stage solution to Michael's "overambitious" Diagram [U1-6]. He had introduced a Metaphor in terms of which solving the triadic addition all at once would amount to bypassing the resources "we" can rely upon. The fact that "we" strive for simplicity, on its own, would not be consequential with regard to the problem

Figure 8.6. Metaphor Mediation Between Diagrams

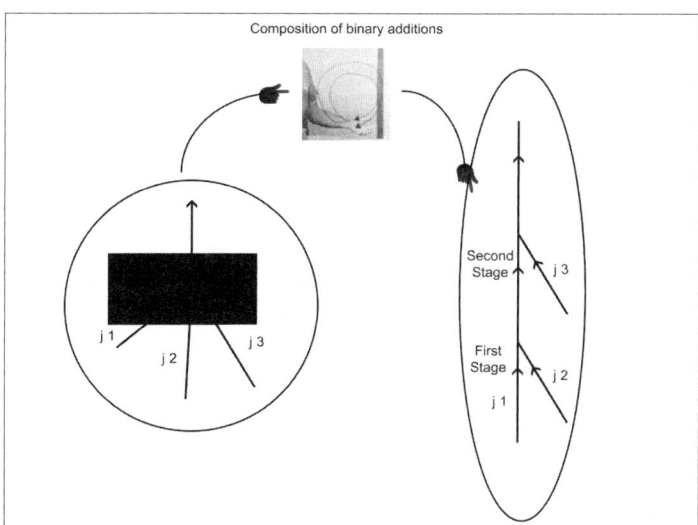

Composition of binary additions

at hand without it being tied to the Diagram under consideration, which makes clear that the power of this Metaphor stemmed from its Diagrammatic expression in the form of binary Diagrams. We suspect that this is generally the case with mathematical Metaphors, in the sense that they are built upon the significance of their Diagrammatic expressions.

Right before Utterance 2, Joseph erased Michael's Diagram and instead drew a new one (see Figure 8.7).

This figure shows a different order of combining the three spins as compared to the first tree Diagram from Figure 8.2. Instead of combining j_1 and j_2 first, the Diagram in Figure 8.7 shows combining j_2 and j_3. Here, the resulting combination is labeled j_5, and the topmost juncture indicates that j_5 and j_1 are to be combined. However, Joseph notices a problem with this labeling scheme, resulting in Utterance 2 (Figure 8.8).

The issue to which Joseph is responding here is that the tree Diagram in Figure 8.2 already specified j_5 to be the result of combining all three spins. For the two tree Diagrams (Figures 8.2 and 8.7) to be describing the same overall process, then, the output of the new tree Diagram had to be labeled j_5 as well, which meant that none of the other edges could be labeled j_5. In [U2-1] he erases the label j_5 and replaces it with a new label, j_6. He then glances at the first Diagram ([U2-2]), keeping his writing hand posed in front of the new Diagram, while he checks that the new and old Diagrams are interrelated properly. Only after doing this does he finish labeling the new Diagram ([U2-3]).

The labels j_n that Joseph added to the Diagrams work as indices because, by means of proximity, they name a particular component that makes it distinctive

Figure 8.7. Second Addition

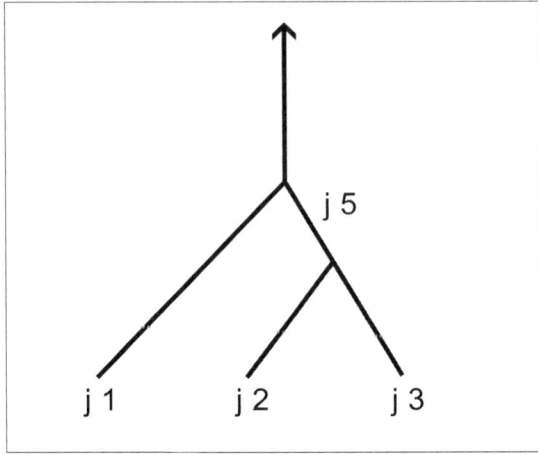

Figure 8.8. Utterance 2

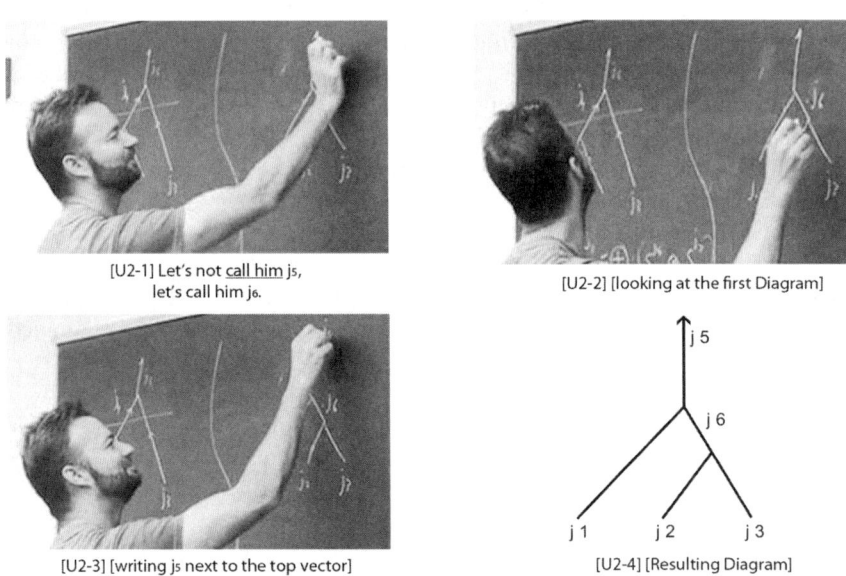

from other ones that otherwise may look exactly the same. What makes these indices indispensable is their power to identify similarities and differences across Diagrams (Netz, 1999). In our daily interactions with objects, we have countless perceptual ways to ascertain similarities and differences. We can easily recognize that, for example, someone's nose as seen from a certain angle is the same as the one seen from another angle. Yet, because Diagrams have a "skeletal" and "bare bones" quality (Stjernfelt, 2000), different Diagrams leave the identity of their components fully ambiguous unless we label them (e.g., is the upper arrow in Figure 8.2 the same as the upper arrow in Figure 8.8, U2-4?). In other words, while the components of a single Diagram are often taken as different just because they are positioned in different parts of the Diagram, something additional is needed to recognize particular components as they appear across Diagrams. Although the vertical vector from the Diagram in Figure 8.2 is the same as the vertical vector from the Diagram in Figure 8.7, the one resulting from $j_1 + j_2$ is different from the resultant of $j_2 + j_3$, and in order to remove ambiguity, Joseph created a new label, j_6, for the latter.

The process of generating the label j_6 was not just a matter of writing it next to that particular edge. To ascertain its location and numerical subindex, Joseph physically navigates the connection across Diagrams by placing his gaze on the first Diagram while keeping his torso and hand engaged with the second one. In a sense, his physically mediated attention brings the two Diagrams (Figures 8.2 and 8.7) into mutual coordination. This observation stresses the significance of treating Diagrams as inseparable from gesture, talk, drawing, body positioning, and so forth (Nemirovsky & Ferrara, 2009). We can describe Joseph's utterance, through his gaze and body posture, as tracing what we refer to as a *Path* from one Diagram to another. Typically, Paths of this sort are embedded in bodily activity and jump from one Diagram to another, preserving relative or partial identities between them. These embodied Paths appear frequently and in many guises throughout the interview with Joseph and with other mathematicians as well.

After Utterance 2, Joseph went on to explain how one could think of the space of possible spins for this three-particle system as a vector space such that the two different ways of combining the three particles corresponds with two different bases e_i and f_j. A basis for a vector space is roughly like axes for a Cartesian plane. Although we usually describe where a point in the plane is located by specifying its horizontal and vertical coordinates (solid axis, Figure 8.9), we could just as well pinpoint it by specifying coordinates on a tilted pair of axis (dashed axes, Figure 8.9).

The point (a, b) can also be described with the "dashed coordinates" (c, d). The "dashed axes" form a different basis for the same Cartesian plane, which means that we describe the points' locations differently. The rule for how to convert coordinates written in one basis to coordinates written in another is commonly described by what is known as a *change-of-basis matrix*. A change-of-basis matrix for the Cartesian plane could tell us how to rewrite "normal" coordinates for a point in terms of "tilted" coordinates, for instance.

Each of the quantum spins j_n occupies its own space, so adding two different spins requires creating a kind of joint space. Adding in the order $(j_1 + j_2) + j_3$ produces a different set of coordinates for the total joint space than does adding in the

Figure 8.9. The Same Point in Different Coordinates

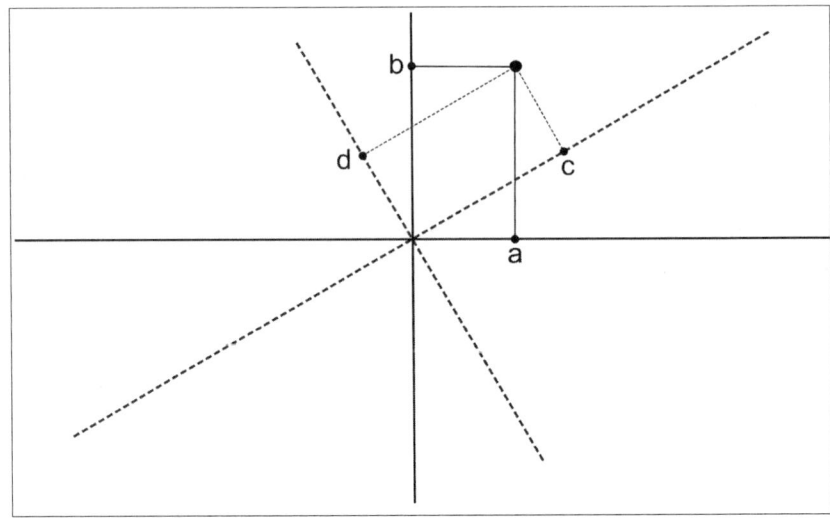

Figure 8.10. Utterance 3

[U3-1] If you're in an inner product space you could think of it really as a matrix

[U3-3] the e_i's

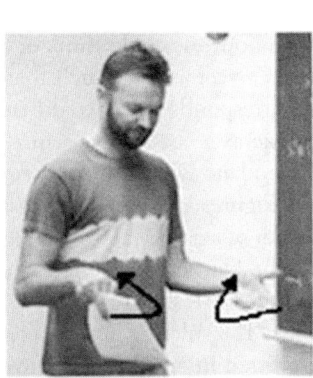

[U3-2]
of <u>inner</u> products
between

[U3-4] and the f_j's

order $j_1 + (j_2 + j_3)$. Thus, some kind of change-of-basis matrix is needed to switch between the coordinates of the two methods of combining three spins.

This leads us to Utterance 3 (see Figure 8.10).

The "it" Joseph refers to in [U3-1] is the change-of-basis matrix from the e_i basis to the f_j one. Notice that Joseph is describing his thinking using the empty space in front of him. At [U3-2], he's staring into a seemingly blank region between his arms, as though looking at something there. His arms seem to indicate the boundary of this empty region into which he is projecting imaginary Diagrams. Right after [U3-2], he looks up and indicates collections (the e_i's to his right in [U3-3], and the f_j's to his left in [U3-4]). The Diagrams here are imaginary but bounded by Joseph's gestures, gaze, and posture as coordinated with his speech, and they are created within the space in front of him accessible to his hands.

The images in [U3-3] and [U3-4] show yet another example of Path-following. The entries of the change-of-basis matrix come from combining (via an inner product) the e_i's and the f_j's. Notice that the Path itself becomes the central object in the space in front of Joseph, with the e_i's and f_j's off to either side. Although we are to understand the two bases as being connected via the gestures of [U3-3] and [U3-4], the entity Joseph seems to want us to attend to is the connection itself. His word choice here might be relevant in this light: He indicates thinking of the matrix as being made up of "the inner products between," not "of," the e_i's and the f_j's.

A crucial aspect for the constitution of Paths across Diagrams is suggested by Joseph's enactment of an imaginary Diagram, which is the need to account for the transformation between them. Imagine that one attempts to constitute a Path between two Diagrams such as $y = 3x-1$ and $xy = 2$. Recognizing that the symbols "y" and "x" are the same in both Diagrams is not sufficient to assert a Path between them because, additionally, we would need to account for what legitimate transformation would produce one given the other. What makes a transformation "legitimate" is that it preserves what is initially "given," which in this case is a certain relationship between "x" and "y." In the example above, there is no legitimate transformation between $y = 3x-1$ and $xy = 2$ because they do not describe the same relationship between "x" and "y." As Joseph was preparing himself to create a new Diagram by merging those depicted in Figures 8.2 and 8.7, in Utterance 3 he tried to articulate what kind of transformation would make such merging legitimate or relation-preserving.

Finally, in Utterance 4, Joseph combines the two tree Diagrams (Figures 8.2 and 8.7) in order to produce the tetrahedron. He does this by first using some arguments from linear algebra to justify drawing the Diagram in Figure 8.11.

The left half of this Diagram represents the first tree Diagram from Figure 8.2 (where j_1 and j_2 were first combined). You can think of the tree Diagram as having been turned 90 degrees counterclockwise and then connecting the j_5 line at the very top (see Figure 8.12). The right half of this circular Diagram represents the second tree Diagram from line [4] of Utterance 2 (where j_2 and j_3 were

Figure 8.11. The Merged Diagram

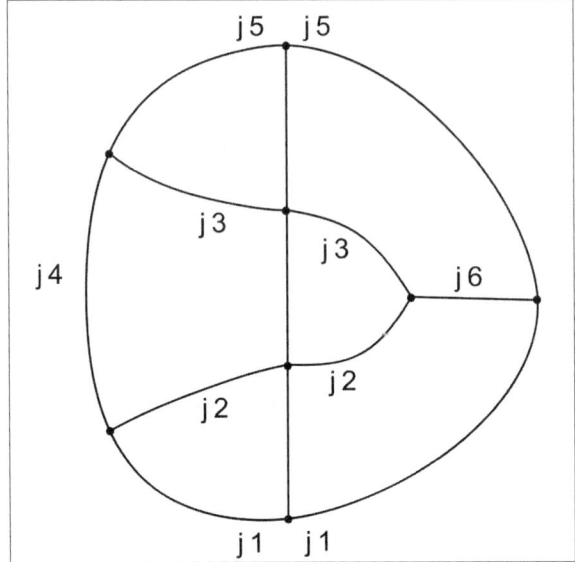

connected first) but turned 90 degrees counterclockwise and mirror-reflected (see Figure 8.12).

The vertical line represents the fact that the inputs (j_1, j_2, and j_3) and the output (j_5) are all the same regardless of how the inputs might be combined. The two halves depict two different ways of combining them, just as the separate tree Diagrams had done before. Notice also that in the original tree Diagrams (Figures 8.2 and 8.7) there were arrows indicating the direction of each combination (e.g., in Figure 8.2, j_1 and j_2 combine to form j_4 rather than j_1 and j_4 combining to form j_2). However, Joseph chose to neglect drawing those arrows as he produced the Diagram in Figure 8.11. This choice was informed by Joseph's familiarity with this type of Diagram-use and, in particular, his anticipation of where he wanted to go next, which requires ignoring the directionality that had been there before. This is at once a critical and nontrivial move: It is essential because without it the symmetries that later emerge would not be there; and it is nontrivial in that the ability to ignore directionality seems to emerge from a subtlety about the algebra involved in combining quantum spins. Indeed, it is still not entirely clear to us as of this writing how to make the case that ignoring directionality constitutes, in this instance, a legitimate transformation. So, here we see again how a Diagram user brings into play a background of life experiences when ascertaining what transformations are both meaningful and useful.

Having created this new Diagram, Joseph reveals the tetrahedron (Figure 8.13).

Diagrams are iconic because they express relations of similarity with the signified object that can be made to appear. They are different in at least two regards

Figure 8.12. Creating the Merged Diagram from Earlier Diagrams

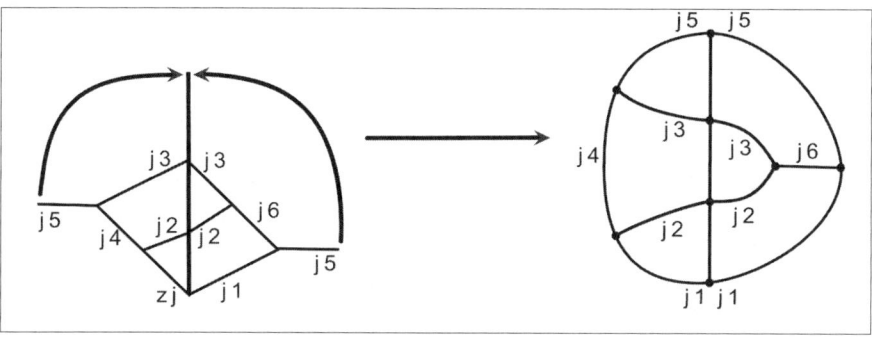

Figure 8.13. Utterance 4

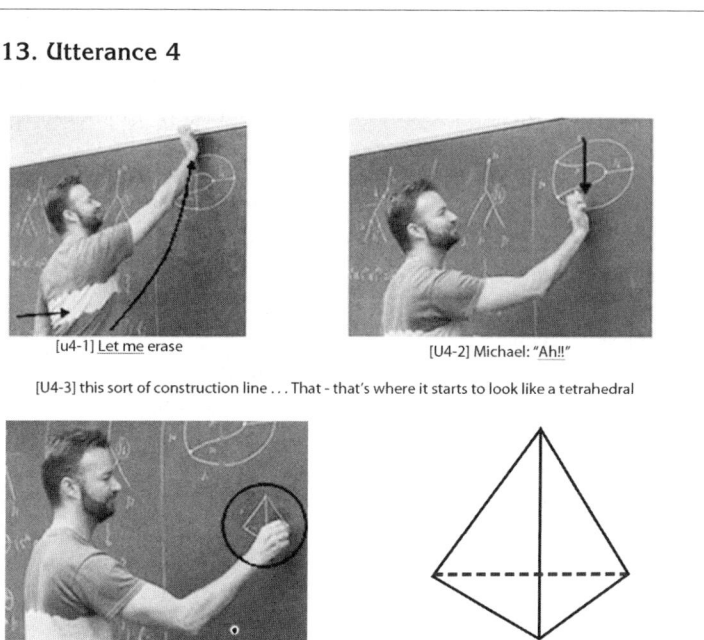

[u4-1] <u>Let me</u> erase

[U4-2] Michael: "<u>Ah!!</u>"

[U4-3] this sort of construction line . . . That - that's where it starts to look like a tetrahedral

[U4-4] sort of arrangement of [pause]

[U4-5] of things [laughs, pointing at the Diagram]

from Images (i.e., those Icons that are similar to the object in their "looks," such as a map of a town or a portrait of a person). One is that Diagrams have a "skeletal" appearance, lacking the richness of detail typical of Images, on which basis Images habitually preserve countless particulars of depicted entities. A second difference is that the transformations we use to manipulate Diagrams are often radically different from the actions we might perform on the signified entities. In contrast, actions on Images tend to reproduce important aspects of the ones we might enact on the signified entities. For example, that we turn our gaze on a map to our left side might serve to identify that we need to turn left on the road.

In Utterance 4, Joseph performs two transformations on the merged Diagram of Figure 8.12; the first one is to erase the vertical segment, just as those who are comfortable with algebraic equations might choose not to write the constant term (i.e., the zero) in y = 2x + 0. Once Joseph does this, something new emerges: We can see the tetrahedron by imagining that we "pull" the middle junction point of the circular Diagram out of the board and then straighten all the edges between the junction points (see Figure 8.14).

Michael's exclamation in [U4-2] expresses this sudden seeing of something new. Joseph's drawing of the tetrahedron in [U4-4] and [U4-5] is then an effort to make the tetrahedron apparent, immediate, using a dotted line ([U4-5]) to indicate the use of a third dimension that extends out of or into the blackboard.

This result is mathematically significant because the regular tetrahedron has a number of symmetries, which is to say that we could rotate it such that the various vertices swap places and the end result would look the same. For instance, we could lift the tetrahedron at its peak, rotate it 120 degrees (i.e., a third of a full rotation), and set it down; the result of having done so will look exactly the same as before. These various symmetries correspond to different ways of looking at the four-spin system (j_1, j_2, j_3, and j_5), including but not limited to the two tree Diagrams Joseph had started with (see Figures 8.2 and 8.7). This, in turn, suggests corresponding symmetries in the various change-of-basis matrices that convert between the different scenarios. Therefore, by doing these Diagrammatic manipulations, Joseph showed us how the tetrahedron's symmetries could be seen in the computations of spin.

Using only relation-preserving transformations, the signified mathematical objects appear to hold object permanence in the sense of a common entity being preserved across Diagrams, which makes it appear as though Joseph revealed the tetrahedron as having always been in the spin computations. Suddenly this feels like a discovery, as though the change in perspective allows one to see something that was already there but was inaccessible beforehand. This is akin to how we might pick up an object like a small sculpture to touch and feel various sides of it, discovering new facets of the object that we assume were there before we started exploring them. We propose that this is part of where the impression of Platonism in mathematics comes from: The permanence of mathematical entities

Figure 8.14. "Pulling" the Tetrahedron from the Merged Diagram

across Diagrams seems to yield "discoveries," which in turn seems to reaffirm their objective reality. We propose this not to argue for or against Platonism, but instead to help illuminate a possible origin for our experience of mathematics as corresponding to real and ideal objects. Even if mathematical objects were Platonic Ideals, the question would still remain as to how we, as embodied non-Ideal beings, are able to explore these mathematical entities at all and to recognize their properties in the course of what we often experience as discoveries. We believe that this embodied connection to object permanence helps answer this question.

DISCUSSION

In this chapter we described Peirce's ideas about mathematical Diagrams, advanced the view of Diagrams as jointly constituted by all types of bodily expression (drawings, gestures, talk, facial expressions, tones of voice, gaze, and so forth), and examined their potential to illuminate an interview episode with Joseph.

We oriented our commentary on Utterance 1 toward the notion of a Peircean Metaphor. We pointed out how the insight expressed by this Metaphor of "simple beings" hinged on its Diagrammatic expression. This view about mathematical Metaphors is different from the one advanced by Lakoff and Núñez (2000), according to which metaphors are unconscious mechanisms mapping a source and a target domain because (1) the subject is conscious of these Metaphors (i.e., Joseph was fully aware of his introduction of a third aspect, namely, a preference to use what we are already familiar with to build up more complex cases, which was not inherent in the use of the tree Diagrams per se); (2) they are explicitly invoked in terms of Diagrammatic expressions; and (3) rather than mapping from a source domain to a target domain, from our perspective Metaphors serve to make the case for the use of certain Diagrams rather than others.

In the context of Utterance 2, we discussed the significance of labeling Diagrams and how labels are indispensable to identify identical components across Diagrams. Furthermore, we pointed out how this labeling entails a kind of bodily traversing across and between Diagrams as expressed through hand motions, posture, gaze displacement, and so forth, which allows for the constitution of Paths across Diagrams. Our reflections on Utterance 3 included commentaries on the notion of imaginary Diagrams populating the empty space within arm's reach of the Diagram user. The particular imaginary Diagram that Joseph placed in front of him led us to consider transformations across Diagrams and the importance of these transformations being legitimate in the sense that they preserve the relationships given in the original Diagrams. Within mathematical practices, only those Paths that traverse Diagrams via relation-preserving transformations are acceptable.

We analyzed in Utterance 4 how traversing a Path can be experienced as a discovery of aspects or properties that seemed present but unavailable to our inspection at the start of the Path. During line [U4-3], Joseph began to draw the image of a typical tetrahedron in three-dimensional space. This tetrahedron emerged from combining the two tree Diagrams into one, twisting the edges around while preserving the topological properties of the overall shape, and dropping the use of arrows indicating directionality along each edge. Because all the transformations had been legitimate, the Path appeared to make visible a tetrahedron that seemed to have already been implicitly part of the initial Diagrams.

To gather together these reflections, we want to highlight two implications that seem to us highly relevant for a theory of mathematical cognition and for learning mathematics. The first one concerns our relationship to mathematical objects. It is customary to assert that "advanced mathematical constructs are totally inaccessible to our senses—they can only be seen with our mind's eyes" (Sfard, 1991, p. 3). The perspective advanced in this chapter, however, suggests that the use of Diagrams allows for *indirect* perceptuo-motor access to mathematical objects: Through iconic mediation, we can physically manipulate and observe them in ways that seem to display some properties and patterns that are not subject to decisions and desires originating in us. This is similar to how, say, it is the shape of a ball that makes it look round to us from all angles. While indirect, Diagrammatic access to mathematical objects can have a degree of tangibility and perceptual appearance that is lacking in the case of many other abstract nouns, such as "justice" or "independence," for which we have symbolic but not iconic means of grappling with them.

The second implication is the opening of a different perspective on the ways in which we constitute mathematical objects, or "reification." Sfard (1991) proposed an influential theory in which she identified three stages: (1) interiorization, through which algorithms, procedures, or, more generally, "processes" become mental entities; (2) condensation, that is, the compression of the process into a compact unity; and (3) reification:

> Only when a person becomes capable of conceiving the notion as a fully-fledged object, we shall say that the concept has been reified. Reification, therefore, is defined as an ontological shift—a sudden ability to see something familiar in a totally new light. Thus, whereas interiorization and condensation are gradual, quantitative rather than qualitative changes, reification is an instantaneous quantum leap: a process solidifies into object, into a static structure. (Sfard, 1994, pp. 19–20)

The alternative that we want to propose is to see reification as embedded in the active and open traversing of Paths across Diagrams. Rather than contrasting static objects and dynamic processes, we suggest that reification emerges from within Path-making practices. As such, reification does not appear to be an "instantaneous quantum leap" but a gradual familiarization with a place of sorts

(Nemirovsky, 2005), populated by certain Diagrams. Our sense of objecthood for a physical entity is similarly grounded in perceptuo-motor use, and it evolves through a lifelong stream of experiences.

ACKNOWLEDGMENTS

This research has been supported in part by the "Tangible Math" project funded by the National Science Foundation, Grant DRL-0816406. All opinions and analysis expressed herein are those of the authors and do not necessarily represent the position or policies of the funding agencies. The authors wish to thank the following people for the feedback received on a previous version by Martha Alibali, Bárbara M. Brizuela, Elizabeth de Freitas, Christian Greiffenhagen, Brian Gravel, Rogers Hall, Kevin Leander, Mitchell Nathan, Nathalie Sinclair, and an anonymous reviewer.

REFERENCES

Bernard, R. (1988). *Research methods in cultural anthropology.* Beverly Hills, CA: Sage Publications.

Campos, D. (2010). Peirce's philosophy of mathematical education: Fostering reasoning abilities for mathematical inquiry. *Studies in Philosophy and Education, 29*(5), 421–439.

Erickson, F. (2004). *Talk and social theory.* Cambridge, UK: Polity Press.

Lakoff, G., & Núñez, R. E. (2000). *Where mathematics comes from: How the embodied mind brings mathematics into being.* New York: Basic Books.

Lehrer, R., Strom, D., & Confrey, J. (2002). Grounding metaphors and inscriptional resonance: Children's emerging understanding of mathematical similarity. *Cognition and Instruction, 20*(3), 359–398.

Nemirovsky, R. (1994). On ways of symbolizing: The case of Laura and the velocity sign. *The Journal of Mathematical Behavior, 13*, 389–422.

Nemirovsky, R. (2005). Mathematical places. In R. Nemirovsky, B. Warren, A. Rosebery, & J. Solomon (Eds.), *Everyday matters in science and mathematics* (pp. 45–94). Mahwah, NJ: Lawrence Erlbaum Associates.

Nemirovsky, R., & Ferrara, F. (2009). Mathematical imagination and embodied cognition. *Educational Studies of Mathematics, 70*(2), 159–174.

Netz, R. (1999). *The shaping of deduction in Greek mathematics.* Cambridge, UK: Cambridge University Press.

Noble, T., Nemirovsky, R., DiMattia, C., & Wright, T. (2004). On learning to see. *International Journal of Computers for Mathematical Learning, 9*(2), 109–167.

Peirce, C. S. (1885). On the algebra of logic: A contribution to the philosophy of notation. *American Journal of Mathematics, 7*(2), 180–196.

Peirce, C. S. (1931–1935). The icon, symbol, and index. In C. Hartshorne & P. Weiss (Eds.), *The collected papers of Charles Sanders Peirce* (Vol. 2, Chapter 3). Cambridge, MA: Harvard University Press.

Peirce, C. S. (1991). *Peirce on signs*. Chapel Hill, NC: University of North Carolina Press.

Sfard, A. (1991). On the dual nature of mathematical conceptions: Reflections on processes and objects as different sides of the same coin. *Educational Studies of Mathematics, 22*, 1–36.

Sfard, A. (1994). Reification as the birth of metaphor. *For the Learning of Mathematics, 14*(1), 44–55.

Stjernfelt, F. (2000). Diagrams as centerpiece of a Peircean epistemology. *Transactions of the Charles S. Peirce Society, 36*(3), 357–384.

Stjernfelt, F. (2007). *Diagrammatology: An investigation on the borderlines of phenomenology, ontology, and semiotics*. Dordrecht, Netherlands: Springer.

CHAPTER 9

Using Representations to Reason About Air and Particles

Brian E. Gravel, Nora Scheuer,

and Bárbara M. Brizuela

Piaget's (1952) constructivist argument is that children have ideas that constitute the genesis of later understandings and knowledge (Piaget & Inhelder, 1969). Within science education, a more recent perspective proposes that students possess vast stores of resources that they use to make sense of new situations and natural phenomena that they encounter (see Hammer, 1996, 2000; Hammer, Elby, Sherr, & Redish, 2005; Rosebery, Warren, Ballenger, & Ogonowski, 2005). This *resources perspective* builds on the central thesis of Piaget's work to posit that students use a variety of ideas, epistemological beliefs, and personal experiences to understand the world around them. The diverse ways in which children put to use their various, personal resources as they attempt to make sense of experiences involving physical and cultural dimensions may give rise to paths in their learning that are inextricably individual as well as cultural in nature. Individual paths stand strongly on cultural tools and constraints, but they cannot be generalized or considered to be true for every other individual. In that sense, we find it helpful to connect the resources perspective with the concept of "learning trajectories" (Sarama & Clements, 2009; Wiser, Smith, Doubler, & Asbell-Clarke, 2009), which underscores this nonlinear quality of child learning in particular contexts.

Central to accessing and using resources are the representational forms the child produces; trajectories are formed through the production, critique, and refinement of these forms. In other words, from a learning standpoint, external representations are the building blocks for students' resources, and they serve to *amplify reasoning*.

Prior work emphasizing the value of students' conceptual resources has focused mainly on students' ways of reasoning verbally about the phenomena they encounter. These studies have shown the generative power of students' intuitive ideas (Hammer, 1995, 2004; Hammer et al., 2005) and everyday language (Rose-bery et al., 2005). We conjecture that in addition to students' talking, other kinds of representations they generate, interact with, and interpret (e.g., drawings, animations, physical artifacts) are both critical instantiations of their ideas *and* essential tools for furthering their thinking and reasoning about science. However, representations (other than verbalizations) have not traditionally been the explicit focus of these works (see Hammer, 2004).

This chapter explores one student's production of various types of external representations and how these artifacts themselves are integral in shaping her learning trajectory as she seeks to generate coherent, increasingly applicable, and robust causal explanations. These kinds of explanations capture the underlying causal mechanisms (Russ, Sherr, Hammer, & Mikeska, 2008) for a specific phenomenon, and they are considered to be central to meaningful science education (Russ, Sherr, Hammer, & Mikeska, 2008).

We focus on the student's explanations—supported by a variety of external representations—of one of those "entities that are not directly observable" (Chinn & Malhotra, 2002, p. 186): air. The reason for this choice is that the use of external representations is critical to render causal patterns of unseen entities explicit; the external representations become the objects a student uses to reason about mechanisms.

REASONING WITH REPRESENTATIONS

The cognitive power of external representations has been highlighted from different standpoints, mainly relating to the contexts of cultural evolution (Donald, 1991; Olson, 1994), human cognitive development (Karmiloff-Smith, 1992), and education (Martí, 2003; Pérez Echeverría & Scheuer, 2009). Karmiloff-Smith (1992) proposed the concept of Representational Redescription, in which she argued that appropriating and using external representations enhances a "specifically human way" to construct knowledge. This representational approach consists of utilizing available or previous knowledge (procedural, conceptual) "by iteratively re-representing in different representational formats what its internal representations represent" (Karmiloff-Smith, 1992, p. 15). Producing an external representation (e.g., a drawing) entails students' selection of some of the many features they perceive, know, or hypothesize about the represented entity. In order to capture these internalized (we prefer this term over Karmiloff-Smith's "internal") representations in an external form, students must decide how to convey such features and how to express them. Besides allowing students to de- and re-compose the represented object (and thus helping them redescribe their ways of understanding or conceiving it), the production of an external representation with relatively endurable materials (e.g., paper and pencil, modeling clay, computer software) leads to a more or less lasting product, which may be seen, commented on, critiqued, stored, retrieved, and even edited during and after the production process, by the person who produced it, by others, and with others (Martí, 2003).

In addition to the productions themselves, the roles of representational activities in promoting reasoning about mechanisms deserves attention. While diSessa's Meta-Representational Competence (see diSessa, Hammer, Sherin, & Kolpakowski, 1991; diSessa & Sherin, 2000) framework focuses on students as competent producers and critics of representations, we seek to extend these ideas to suggest that external representations produced with some intent or goal—either communicative or epistemic—capture in particular contextual ways some of the students' sense-making resources, amplifying cognition with regard to understanding the natural (and social) world. Vygotsky (1981) argued that "signs" (i.e., representations) mediate the child's growth of logical thinking and reflection.

Considering students' competencies for representation from the standpoint of multiple representations strengthens this position. According to Pérez Echeverría and Scheuer (2009), "the use of alternative external representations to describe a single situation assists the explication of epistemic attitude, across developmental periods, learning situations and domains of knowledge" (p. 11). Other researchers have made a similar argument, where attempts to express ideas across systems of representation are shown to be beneficial for developing understanding (Brizuela & Earnest, 2008; Gravel, 2011; Lehrer & Schauble, 2002). One specific benefit of using and producing multiple representations lies in the possibility that each system of representation may highlight aspects of a problem that others do not (Kaput, 1998; Pérez Echeverría & Scheuer, 2009; Zhang & Norman, 1994). That particular conceptual features are made more salient in certain systems, and that specific types of representations support specific types of reasoning, are both stances that are supported by the literature in mathematics and science education, among other fields.

Let us note that this range of cognitive gains is particularly pertinent in situations when students create an external representation for an object they are trying to understand and when no conventional external representation is available. An example of this type of scenario is elementary school students' response to a request to produce external representations for an unseen such as air, as we will describe in this chapter. Given the theoretical arguments and empirical evidence in favor of multiple representations in the exploration of both new and familiar phenomena, the lens for this chapter is the production and critique of external representations of an "unseen."

EXPLORING CHILDREN'S REPRESENTATIONS OF "EXPERIENTIAL UNSEENS"

There are many aspects of the natural world that we feel or experience, but cannot see. We assign the term *"experiential unseens"* to these phenomena, such as atoms, molecules, or sound; this chapter explores air as an "unseen." Air is an incredibly commonplace, ubiquitous, and multidimensional idea present in students' lives from very early ages (Driver, Squires, Rushworth, & Wood-Robinson, 1994; Piaget, 1930/2001). Researchers have pointed to students' strong associations of air with movements caused by wind, such as a breeze through a tree's leaves, and have asserted that elementary school children only believe in the existence of air in the context of some process of change (e.g., wind currents; see Driver et al., 1994, for a review). Alternatively, an emphasis on the particle nature of matter specific to air and gases has been shown to support students' efforts to understand the structure of air as a substance (see Johnson, 1998; Nussbaum & Novick, 1982). While some researchers have characterized students' understandings by their degree of "accuracy" (e.g., Lee, Eichinger, Anderson, Berkheimer, & Blakeslee, 1993), focusing on what students either know or do not know is at odds with the resources perspective and with the focus of this chapter. Instead of describing students' shortcomings or limitations, in this study we focus on exploring *how* students use representations to reason about mechanisms in the context of air.

In order to work with the "unseen," students must name, talk about, and make "visible" (or, as we will argue, represent) processes and mechanisms that are not directly observable. From this perspective, we present a study guided by the research question: How do students produce and critique multiple representations to make sense of and explain air and the particulate nature of gases?

IRIS'S REPRESENTATIONS OF AIR

We examined how sense-making resources are accessed through representations, and how this *amplifies* reasoning about mechanisms to describe cause and effect by interviewing 12 5th-grade students at an urban public school in the Greater Boston area. The interviews focused on a linked syringe device (see Figure 9.1), through which students explored the compressible nature of the gas and how air transmits forces. Students were requested to explain the behavior of the syringes through their verbalizations, drawings, and by creating an animation (specific prompts and requests are included below).

To further explore how students accounted for the physical processes involved, each interview session concluded with the students critiquing a prepared representation. In addition, students were asked to compare their representation of air to linked syringes containing water (in order to probe their ideas about the material nature of air and water relative to their compressibility).

We present the trajectory of Iris as an interesting case showing how a 5th-grade student accesses and uses her cognitive and representational resources in particular contexts to construct coherent explanations of air.

Iris, at the time of the interviews, was a 10-year-old African American student. Her teachers described her as curious and articulate, and said that she enjoyed doing science—experimenting, exploring, and "messing about" with explanations

Figure 9.1. Linked Syringes Used in Interviews. Students Could Either Push One Syringe Plunger (the Pushing Case) or Push Both Plungers Simultaneously (the Compression Case).

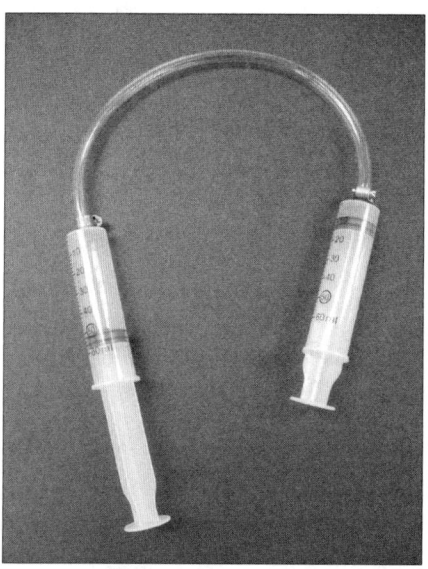

and ideas. This tendency, coupled with her affinity for scientific terminology, makes her case study interesting and illustrative.

We follow Iris across two interview sessions, occurring on consecutive mornings, by combining the presentation of selected fragments of her verbal explanations and her productions of external representations of air (included in the tables below) with our narrative, which offers a reconstructed and interpretive account of her evolving reasoning and representational processes. We present the results in this manner to provide the reader with options for engaging with the ideas: through Iris's verbalizations and representations, and through our interpretation and analysis of her work.

"What Do You Know About Air?":
Iris's Verbal and Drawn Representations

When asked what she knew about air, Iris said that it was a gas, made of "oxygen stuff," that humans need air to live and breathe, and that "air is everywhere." Like all of the students in this study, Iris's examples of air were related to wind and memories of wind moving things such as her lunchbox, which the wind had blown off a ledge earlier in the week. Her initial ways of talking about air combined her previous experiences interacting with the substance with the use of more canonical terminology (e.g., using the word "*oxygen*").

Figure 9.2. Iris's Initial Verbal Statements and First Drawing

Utterances are labeled "u*n*" with *n* corresponding to the number of the selected utterance. Drawings are labeled in the same way, but using "d" for drawing (d*n*).

Initial verbal explanations (prior to request to draw)

"Yeah, because all the air from here goes through the tube and into here, and it's just like the same amount of air bounces back and forth [between the two syringes]." (u1)

"Because all the air, like, all the air doesn't fit in it [the syringes and connecting tube]. So now they have to push back." (u2)

"Because all the air from both of these [the syringes] go into the tube, and then when you have your stopping point, it's because all the air, like there's too many. There's too much air in the tube [connecting tube]." (u3)

First drawing (d1): "Put something on paper that shows me what you know about air and the linked syringes."

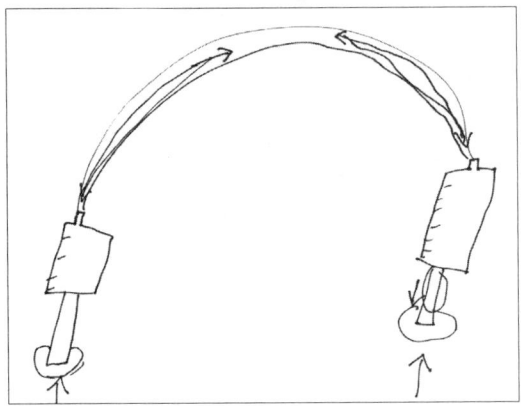

"All the air goes into here [connecting tube], and then right when all the air meets, it has to bounce back, and then that's [the "bounce back"] what pushes this [the plungers] back down." (u4)

When asked to explain her drawing (d1), Iris used arrows and circling to elaborate her explanation:

- Arrows pointing upward on syringe plungers indicate "we're pushing both of them" (u5);
- Arrow pointing downward next to circled plunger handle showing "and that's what pushes this [the plunger] back down." (u6)

When asked to explain what happened "when all the air meets":

"Well, you can't really see air." (u7)

When Iris was presented with the compression case in the linked syringe device—where she could push the plungers to a certain point where air's resistance to compression caused them to stop—her reasoning hinged on the quantity of air ("there's too many"; see u3, Figure 9.2) and a lack of space ("all the air doesn't fit"; see u2, Figure 9.2) in the connecting tube. Her explanation included an implicit idea that each syringe contained a fixed amount of air, and that the combined volumes ("all the air"; see u1–u3, Figure 9.2) were "too much" to fit in the tube connecting the two syringes. Iris also suggested that there is a constant quantity of air ("the same amount"; see u1, Figure 9.2) in the linked syringes; her answers about the compression case point toward a model of two quantities occupying and filling a fixed space (i.e., the connecting tube).

Following the discussion of compression, Iris was asked to put something on paper explaining her reasoning about air; she decided to focus on drawing the compression case first (as opposed to the pushing case). Iris's explanation of compression as two quantities of air meeting to generate the "bounce back" (see u1 and u4, Figure 9.2) became part of her drawn representation (see d1, Figure 9.2), as well as of her verbal explanation of what she produced.

Iris now offered a more detailed description of compression when explaining her drawing. She added arrows and lines to her drawing in different ways to highlight elements of the compression case and to show the motion of air (see d1, Figure 9.2) concurrently with her verbal descriptions. Just as language helped Iris organize her verbal explanations before the drawing task (see u1–u3, Figure 9.2), her drawn representation helped her further organize elements of the situation. Her verbalizations clarified aspects of her drawing, which ultimately made explicit the major components of her explanation: two quantities of air, the lack of space in the tube for the two quantities, and a "bounce back" once the quantities have "met in the middle."

Iris was then asked what happened "in the middle," where she believed the air "met," a space she left blank in her drawing (see u4, Figure 9.2). Iris stated: "You can't really see air" (see u7, Figure 9.2); this response, while interesting, is relatively predictable, given that asking students to draw an "unseen" is a challenging task. Drawing an unseen contradicts one of the main principles of figurative drawing: iconicity (Peirce, 1931–1958). Her claim excused her from having to choose (or invent) a means for representing compression on paper. The lines in her drawing (see d1, Figure 9.2) appeared to represent the movement of air from each syringe into the connecting tube (see u4, Figure 9.2), but she did not explicitly claim that the lines represented air as a substance. At this time, Iris's use of a symbol for air as a substance and a means for capturing compression on paper were in flux.

To probe her reasoning and her representations of air, she was asked to show on paper what air would look like in a closed container. After Iris suggested using colored water (i.e., something that would be visible, see u8, Figure 9.3) to

Figure 9.3. Iris's Second Drawing and Verbalizations

Air in a box drawing task (d2): "How could you show that there is air in this container?"

"Maybe you could like take colored water instead of air." (u8)

Iris writes the word "*air*" inside what looks like a cloud and says, "Just in case they [the viewer] didn't know."(u9)

represent air, she settled on drawing a cloud-like symbol (see d2, Figure 9.3) while continuing to express discomfort with drawing an unseen. Iris said she was concerned that someone viewing her drawing would not know what her idiosyncratic inscriptions meant (see u9, Figure 9.3), which led her to subsequently write the word "*air*" in her drawing (see d2, Figure 9.3)—her first explicit focus on the communicative role of representations. However, when she was presented with the *dots on paper* representation of air (see p1, Figure 9.4), she resorted to the use of a canonical term: "*molecules*" (see u10, Figure 9.4). We wonder whether her descriptions of molecules moving (see u11, Figure 9.4) were a phrase she was repeating (quite adequately) after having heard it elsewhere. While she recognized dots as molecules, Iris's resistance to the presented representation seemed to be fueled by her communicative commitments and her theory of mind (Wellman, 1990) attributions; other children (especially if they are young and in the early school years) may not perceive dots as standing for air molecules (see u12, Figure 9.4). The dots did not present Iris with an explicit advantage over writing the word "*air*" to show the presence of air to a potential, or internalized, audience.

When asked to redraw the compression case (see d3, Figure 9.5), despite now being aware of the *dots* representation, Iris continued to use lines to show air moving in the syringes. A possible interpretation for this is that a line depicts how a substance (i.e., air) moves from one point to another and dots were no more effective for communicating air's motion than lines. This time, she added written labels, presumably because she had imagined an audience for her representation,

Figure 9.4. Iris's Response to a Presented Representation (p1)

Presented "dots on paper" representation (p1)

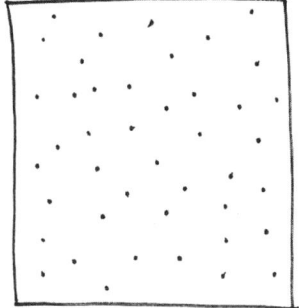

"Oh yeah! Because of the ... um ... 'cause in science we were talking about ... I forgot what they were called, I think they're called molecules?" (u10)

"When you have it in, like, ice, they don't move around a lot. When you have it in water, they kind of move around a lot, and when you have it in a gas they move around a lot." (u11)

Brian: Are dots a good way to show air?

Iris: No ... I don't know, because like, it's just dots. What if like a kindergartner came up? They wouldn't know what the dots meant. (u12)

and because she wanted to make the implicit ideas contained within her production explicit to the viewer. After drawing the syringes (see d3, Figure 9.5), Iris elaborated on the idea of air not being able to "fit" into the connecting tube by formulating a "cup and water" analogy (see u15, Figure 9.5), which served to partially clarify her description of the compression case by giving her something explicit to compare with air. Earlier in the session, Iris mentioned that one could not see air and she suggested using water as a means for showing air (see u8, Figure 9.3). The unseen nature of air posed Iris with a deeper challenge than simply inventing an inscription to stand for air on paper; it also required reasoning about a way to conceive of air as a substance.

When Iris was asked about "what happens to the amount of space that's inside here" (the linked syringe system) in the compression case, she acknowledged that the space becomes smaller, and that air is compressed (see u16, Figure 9.5). When asked to explain "compressing," Iris said "molecules" (see u17, Figure 9.5) again, but to make sense of how a volume of air can get smaller, she now referenced the *dots on paper* representation (see p1, Figure 9.4) presented to her earlier. We interpret that her earlier use of the term "*molecules*" (see u9, Figure 9.3), which was prompted by the interviewer's *dots on paper* representation (see p1, Figure 9.4), made the idea of discrete parts of air more accessible to Iris; molecules in the

Figure 9.5. Iris's Third Drawing and Verbalizations

Revised drawing (d3): "Can you create another drawing of the syringes?"

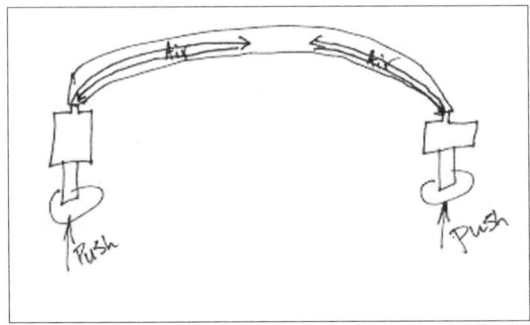

"...when the air from here [coming from the left syringe], meets up with the air from here [coming from the right]...and then when they meet, they have to bounce back to where they came from." (u13)

"The tube [connecting tube] can't hold that much air." (u14)

"'Cause it's just like pressing all this air into something that won't fit it. Just like trying to fit a bunch of water into a small cup." (u15)

Unpacking ideas about compression

Iris: Well . . . you're just compressing it. (u16)

Brian: What do you think if we could zoom way in, what would it look like to compress air?

Iris: Think it would look like all the molecules coming together. . . . Kind of like how the 4th-grader did it [using dots reference, having seen p1]. (u17)

context of thinking about changing volumes helped her think about a potential mechanism for compression (see u17, Figure 9.5).

This new context (i.e., volumes) provided Iris with a bridge between her ideas about quantity and a discrete representation of air. In this way, she began to conceive of molecules (or "dots") coming together when air is being compressed—an idea she then integrated in her drawings of molecules in compressed (see d5, Figure 9.6) and ambient air (see d4, Figure 9.6).

Iris's dot-based drawings (see d4 and d5, Figure 9.6), together with her verbal explanations of molecules spaced differently under compression and "when we're not pushing" (see u18–u20, Figure 9.6), are a compilation of many ideas working in concert to explain the behavior of air in the syringes. Iris referred explicitly to molecules when the graphic representation of air with dots was introduced to her (see p1, Figure 9.4), but did not immediately use that representation. Thinking about compression, specifically, created a need for describing how a volume of air

Figure 9.6. Iris's Appropriation of the Dots Model

Iris's dots representations showing air molecule spacing when air is being compressed (top, d4) and in ambient air (bottom, d5).

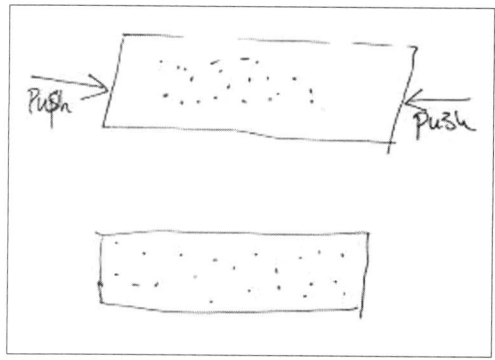

"... and then I imagine they come together ... kind of like with dry ice." (u18)

"Because dry ice changed from a solid to a gas. So, when it's in a solid, all the molecules, they're not moving a lot. But then, once like, it starts getting warm, all the molecules spread out and it becomes a gas." (u19)

Brian: What is air like when we're not pushing (i.e., ambient)?

Iris: They'll just be spread out and all over the place. And, like, they would have their own space. (u20)

Describing compression in the water

"The air, I think, is like more flexible [than water], kind of." (u21)

"'Cause the air ... it's like if you had a flexible person and a not flexible person. They could touch their toes, but the other person couldn't. It's kind of like this one [water-filled syringes] can move just a little bit, but that one [air-filled syringes] can move a lot till it's at its limit." (u22)

in the syringes can decrease. In other words, she realized the cause (compression case) and effect (change in volume), but she needed to develop a mechanism or linking process to connect the two. Iris first reasoned about volume using a specific quantity of water fitting in a cup. We hypothesize that the focus on quantity (refining her ideas of "too much") led Iris to create links between molecules, molecule spacing, and a *dots-on-paper* representation of air to arrive at a sophisticated drawn representation (see d4 and d5, Figure 9.6) and verbal explanation of a mechanism for compression.

Yet, the application of these ideas to water-filled syringes highlighted the limited reach of her *dots* representation and her ability to reason about molecule spacing at this time. Iris introduced the concept of "flexible" to describe the difference between air and water (see u21, Figure 9.6). Iris created an analogy where a flexible person was like air, and an inflexible person was like water (see u22, Figure 9.6); her descriptions of the difference in compressibility between air and water did not involve molecules at this time.

This first session exhibits a complex relationship between the *resources* Iris used to reason about air (i.e., canonical terms, experiences with wind, analogies about flexibility, and quantities of water) and her *representations* (i.e., drawings with arrows and text labels and interpretation and adaptation of a dots model). She demonstrated an ability to change her ideas and her representations to reason about and construct explanations for what she experienced with the linked syringes. At different points, either her graphic representations (or representational criteria—e.g., the communicative intent of a drawing) or her use of canonical terms served as bridges to help her make sense of how air can be compressed. Her agility working with contexts and ways of framing the problem to reason and produce personally satisfying explanations is indicative of the power that production and refinement of ideas and graphic representations can have on attempts to find meaning. As Iris's explanations gained coherence and sophistication in the second interview session, the power of representations became more salient.

Generating a Stop-Motion Animation of Air

The next morning, Iris was charged with making a stop-motion animation of what she knew about air and the linked syringes. She first drew the syringes on a whiteboard with a series of dots ("like the other kids did," recalling again the drawing she was presented with the day before; see p1, Figure 9.4), showing air in the entire length of the tube connecting the two drawn syringes. Next, she erased this and spent 5 minutes creating seven new frames showing a small number of dots moving forward through the tube in each frame (see f1 to f7, Figure 9.7).

Iris seemed to be more comfortable representing an unseen in this session than in the previous session (see u23, Figure 9.8). In the first session, she had somewhat resisted adopting the dots representation for air molecules, instead preferring lines. In this session, she appropriated the model-like nature of dots in her animation. Dots stood for molecules in a general way, but were not meant to stand for each molecule on a one-to-one basis (see u24, Figure 9.8) or to provide an analogical portrait of their size (see u25, Figure 9.8). It seems Iris was aware that she was relying on a large-scale magnifying procedure in order to show the invisible (see u26, Figure 9.8) in her animation.

Following a description of the pushing case, Iris was asked how she might show compression in her animation. Her description mirrored what she said during the first session, with the exception of the phrase "too tight" as a means of describing "all the air" going into the connecting tube (see u27, Figure 9.10). The idea of "too tight" refers both to quantities and relative spacing, which is a

Figure 9.7. The First Seven Frames of Iris's Animation

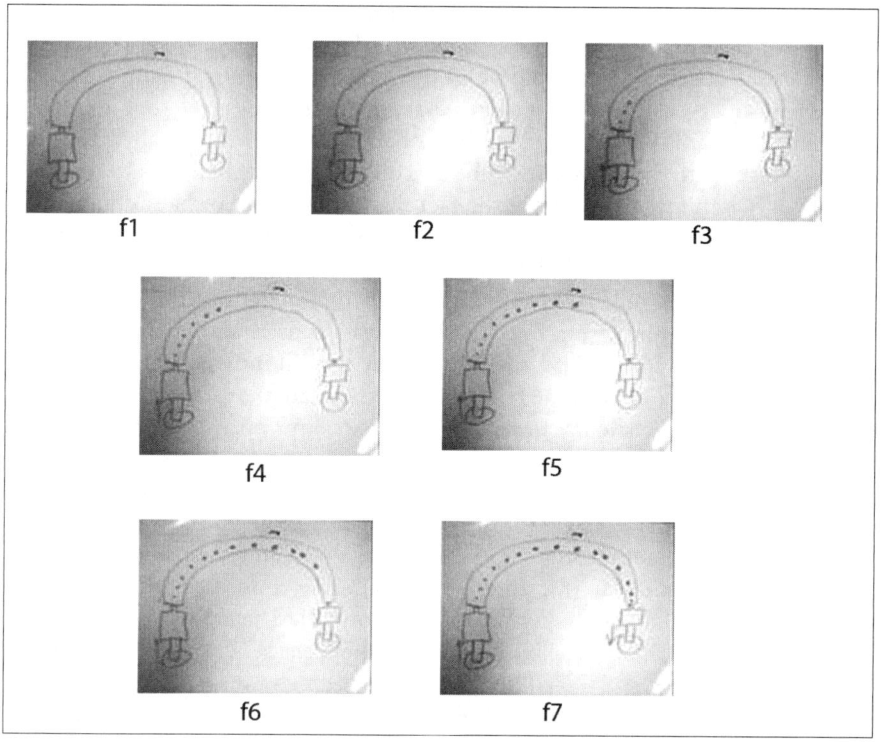

Figure 9.8. Iris's Verbal Descriptions of Her Animation

Brian: These [dots in the animation] are showing the air molecules?

Iris: Yeah, because . . . and I made them a little bit spread out because air is a gas, and gas molecules are really spread out and liquids are kind of spread out, but solids are really close together. (u23)

Brian: So, do you think this is how many air molecules are in there [inside the syringes], or would there be more of them [air molecules]?

Iris: Probably more. (u24)

Brian: Probably more? And do you think they are this big [points to dots in animation], or are they smaller than that?

Iris: Much, much, much smaller. (u25)

Brian: Can we see them?

Iris: No. (u26)

more sophisticated, multidimensional idea than the "too much" idea that she had expressed in the previous session. After offering this description, Iris added six frames to her animation to show a process of dots leaving each syringe frame-by-frame, and meeting in the middle of the tube (see f8—f13, Figure 9.9). Iris's animation showed the "push" from each syringe using arrows (see f8, Figure 9.9), and then it showed two distinct quantities of air moving symmetrically from each of the syringes into the connecting tube. Iris did not show differences in molecule spacing in her animation—despite the fact that she had just verbally commented on them (see u31, Figure 9.10). We can only speculate about the reason for this mismatch between her verbal and her graphically animated representations of the same phenomenon. Did it happen because drawing differences in the concentration of dots is labor intensive, or because she was focused on drawing the direction in which molecules moved and not on how they fit into the limited space of the connecting tube? Or was she first trying to articulate the idea verbally, perhaps in a medium in which she felt more fluent, before committing to a graphical representation?

Iris's depiction of two quantities "meeting" in the middle of the tube was particularly interesting (see f11–f13, Figure 9.9). She made an inscription at the midpoint of the connecting tube in the shape of a star, and she proceeded to add lines with arrows (see f12–f13, Figure 9.9) extending from the star back toward each of the syringes. Finally, she added small downward-facing arrows at each of the syringes (see f13, Figure 9.9) to indicate the "release" (see u28–u30, Figure

Figure 9.9. Iris's Final Six Frames of Her Animation

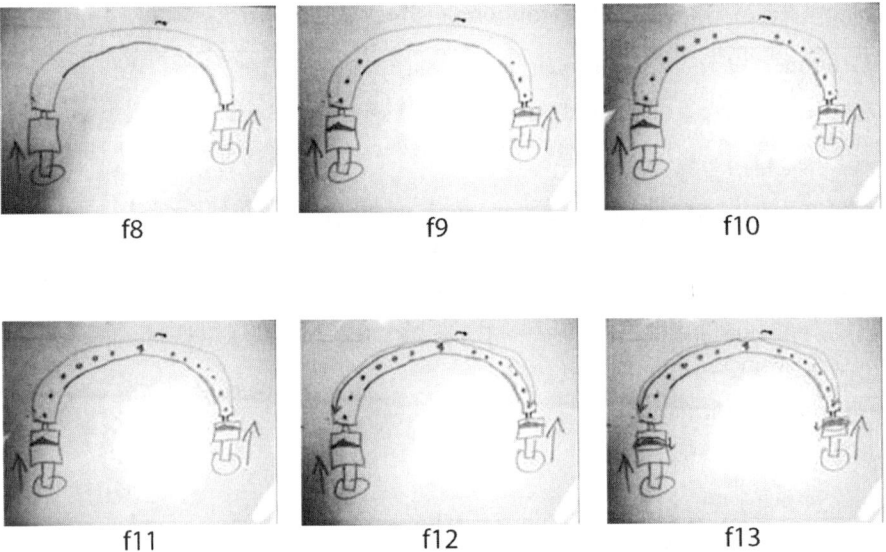

f8   f9   f10

f11   f12   f13

9.10), which pushes the plungers back out of the syringes. Whereas her first uses of ideas about molecules were nascent, she now reasoned about molecule spacing, the size of molecules, and she began constructing an explanation for why the plungers "release" after compression. She evoked the idea that air "needs to go around everything, to make everything live" (see u32, Figure 9.10), which may indicate an attempt to explain the at-rest case, where the syringe plungers are not being pressed. Perhaps Iris was searching for an encompassing explanation, which could account not only for events and changes, but also for states (see Pozo & Gómez Crespo, 1998). An interpretation of Iris's reasoning is that since air is "everywhere," and one of the primary ideas about air is that it helps humans breathe and plants live, then it returns to the at-rest condition because it simply needs to be "everywhere." Her use of a molecule representation (verbally and graphically) served to support her reasoning about molecule spacing and compression, which she was then able to apply to reasoning about the water-filled syringes.

When asked to compare air and water, for the first time in the interviews Iris combined the idea of molecule spacing with her "flexible" construct to offer a more articulate explanation for why water and air have different levels of compressibility (see u33, Figure 9.11). As she explored this idea verbally, she generated another analogy wherein water was like "tight jeans" that are "hard to move in" (see u34, Figure 9.11).

When shown the air-as-dots example animation (see p2, Figure 9.11), Iris believed it showed air, but that the spacing between particles was too tight (see u35, Figure 9.11). Armed with the resources of a discrete model of air and water, a means for describing compression, her own animation, and a ready-made example that she could critique (see p2, Figure 9.11), Iris made significant strides in her reasoning about air, making reference to springs to explain the case of compression of air in the example animation (see u36–u38, Figure 9.11). The integration of these resources resulted in her constructing a mechanism that explained compression, and she reasoned about and related this mechanism for compression to other ideas that she considered to be similar (i.e., springs).

REPRESENTATIONS, RESOURCES, AND REASONING

This case study of Iris demonstrates a complex and dynamic interplay between Iris's resources captured in her verbalizations and productions, in certain contexts, and her developing explanations for a mechanism to describe compression in air. Drawing from a resources perspective, which posits that students use a variety of ideas, epistemological beliefs, personal experiences, and representational tools to understand the world around them, we sought to explore how providing students with opportunities to express their ideas in different ways impacted the explanations they generated for complex phenomena. Reasoning about experiential unseens requires ways of accessing aspects of the phenomena so as to construct relations between

Figure 9.10. Iris's Verbal Explanations of Her Animation

Explaining the compression case

"Okay. So, if you push on this [left-hand syringe] and push on this [right-hand syringe], all the air from there [left-hand syringe] comes over here [into the connecting tube], and all the air from this [right-hand syringe] goes over here [connecting tube], and then when it gets too tight over here [the air in the connecting tube], it has to move back over here [into the syringes]." (u27)

Explaining the star in the middle of the connecting tube (f11)

Iris: That's when it gets to, like, its tightest point, and then it has to release. (u28)

Brian: And the release is what you showed with the arrows going back?

Iris: Yeah. (u29)

"Because when all the air gets too tight in the middle, it has to find a way to exit out. And . . . since these are all the way up here, it has to push itself down to make room." (u30)

Frame 12—She adds lines to show air coming from that "tightest point" back to the syringes.

Explaining "too tight"

Iris: Like, it's all [the air] coming close and close and close together. (u31)

Brian: Ok . . . so, do you think these dots are actually getting closer together?

Iris: Maybe . . . If you have too much air in one thing, it'll pop! So, it's kind of like this [the compression case]. When it [the linked syringes] had too much air in the tube [connecting tube], you had to release it. Because . . . air needs to go around everything, to make everything live. (u32)

causes and effects. In other words, multiple representations are particularly helpful in order to unpack complicated phenomena (such as compression in air) in order to construct detailed explanations of mechanisms. Iris's trajectory presents one case that highlights central features of exploring scientific phenomena using a multirepresentational approach. Iris generated external representations that (1) supported the construction of these explanations, (2) motivated her to generate analogies (e.g., "flexible," "jeans," "springs") and other means of describing her ideas, and (3) ultimately served as amplifiers of her ideas that supported her reasoning about air. As other researchers have argued, students can build from their resources to begin reasoning about causal mechanisms. Iris's case extends this argument by demonstrating how representations contain certain resources in particular contexts, and how these representations can amplify reasoning about mechanisms.

Through the sessions, Iris moved from a focus on quantity, to the formation of a two-quantity model of air in the syringes, and eventually to thinking about how a volume of air can be decreased. Her commitment to the communicative

Figure 9.11. Iris's Comparison of Water and Air and Her Critique of the Example Animation

---

Describing the water-filled syringes

"Because . . . water is not as flexible as air, because all the molecules are closer together." (u33)

"What I mean by flexible is like, 'cause gas is all the molecules like floating around and they've got their own space, and the water, it's not as spacey as that. So, it's kind of like, more packed. Like, it's kind of like if you wear tight jeans. It's kind of hard to move in them. But if you wear your size jeans, it's easy to move around." (u34)

Critique of example animation (p2)

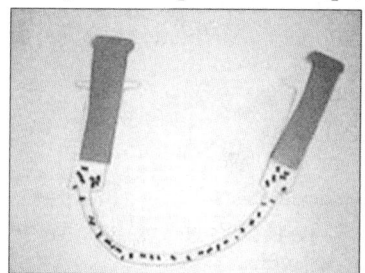

Iris: She's [the assumed creator of the example animation] showing the air by the molecules I think again, but she kind of made them a little bit tight, like they're more spread out. I think that it would kind of be a better way to show water. (u35)

Brian: And what happens to all the little molecules of water or air?

Iris: They get tighter and tighter and tighter. I think that's the same thing that I was saying, because they get all close together, and then they have to spread out again. (u36)

Brian: Yeah, what's making the air spread out again?

Iris: Because . . . it's kind of like a spring. (u37)

Brian: Okay, tell me a little more.

Iris: 'Cause if you put it [the spring in relation to air] all close together, and then you let go, it bounces back to where it was before. (u38)

---

aspects of drawn symbols led her to resist the appropriation of "dots," yet the discrete representation of air afforded her a means of articulating a possible mechanism for compression—the dots getting closer together. Once she had proposed this mechanism, Iris was able to further explore and refine her descriptions of compression—using symbols, analogies, and comparisons between air and water—to ultimately construct a complex explanation. It is our contention that Iris's proposal of a mechanism for compression was the result of a sequence of representational productions, critiques, and refinements—or an individual trajectory—that supported and amplified her ability to construct a coherent explanation for compression.

More generally, our study illustrates how activities that build on students' resources through the production and interpretation of multiple representations present opportunities for enriching science instruction. The design of the exploration and representational tasks (both productions and critiques) in this study were meant to support students' efforts to build on their ways of thinking. Perhaps experiential unseens present a unique opportunity not only to study how students use multiple representations, but also for the design of meaningful experiences where students can unpack familiar, yet complex, ideas in multirepresentational ways. If representations provide access to resources and amplify cognition, as we argue, then multiple, iterated interactions with external representations should be emphasized both in educational research and in classroom contexts.

ACKNOWLEDGMENTS

The second author conducted this study with support from Universidad Nacional del Comahue (B-139) and Consejo Nacional de Investigaciones Científicas y Técnicas (CONICET) (PIP 1029) in Argentina.

REFERENCES

Brizuela, B. M., & Earnest, D. (2008). Multiple notational systems and algebraic understandings: The case of the "Best Deal" problem. In J. Kaput, D. Carraher, & M. Blanton (Eds.), *Algebra in the early grades* (pp. 273–301). Mahwah, NJ: Lawrence Erlbaum Associates.

Chinn, C. A., & Malhotra, B. A. (2002). Epistemologically authentic inquiry in schools: A theoretical framework for evaluating inquiry tasks. *Science Education, 86*(2), 175–218.

diSessa, A. A., Hammer, D., Sherin, B., & Kolpakowski, T. (1991). Inventing graphing: Meta-representational expertise in children. *Journal of Mathematical Behavior, 10*, 117–160.

diSessa, A. A., & Sherin, B. L. (2000). Meta-representation: An introduction. *Journal of Mathematical Behavior, 19*, 385–398.

Donald, M. (1991). *Origins of the modern mind*. Cambridge, MA: Harvard University Press.

Driver, R., Squires, A., Rushworth, P., & Wood-Robinson, V. (1994). *Making sense of secondary science: Research into children's ideas*. London, UK: Routledge Falmer.

Gravel, B. E. (2011). *Elementary students' multiple representations of their ideas about air*. Unpublished doctoral dissertation. Medford, MA: Tufts University.

Hammer, D. (1995). Student inquiry in a physics discussion. *Cognition and Instruction, 13*(3), 401–430.

Hammer, D. (1996). Miconceptions or p-prims: How may alternative perspectives of cognitive structure influence instructional perceptions and intentions? *The Journal of the Learning Sciences, 5*(2), 97–127.

Hammer, D. (2000). Student resources for learning introductory physics. *American Journal of Physics, Physics Education Research Supplement, 68*, 552–559.

Hammer, D. (2004). The variability of student reasoning, lecture 1: Case studies of children's inquiries. In E. Redish & M. Vincentini (Eds.), *Enrico Fermi Summer School, Course CLVI* (pp. 279–299). Bologna, Italy: Italian Physical Society.

Hammer, D., Elby, A., Scherr, R. E., & Redish, E. F. (2005). Resources, framing, and transfer. In J. P. Mestre (Ed.), *Transfer of learning from a modern multidisciplinary perspective* (pp. 89–120). Greenwich, CT: Information Age Publishing.

Johnson, P. M. (1998). Progression in children's understanding of a "basic" particle theory: A longitudinal study. *International Journal of Science Education, 20*(4), 393–412.

Kaput, J. (1998). Representations, inscriptions, descriptions and learning: A kaleidoscope of windows. *Journal of Mathematical Behavior, 17*(2), 265–281.

Karmiloff-Smith, A. (1992). *Beyond modularity.* Cambridge, MA: MIT Press.

Lee, O., Eichinger, D. C., Anderson, C. W., Berkheimer, G. D., & Blakeslee, T. D. (1993). Changing middle school students' conceptions about matter and molecules. *Journal of Research in Science Teaching, 30*(3), 249–270.

Lehrer, R., & Schauble, L. (2002). Symbolic communication in mathematics and science: Co-constructing inscription and thought. In E. D. Amsel & J. Byrnes (Eds.), *Language, literacy, and cognitive development: The development and consequences of symbolic communication* (pp. 167–192). Mahwah, NJ: Lawrence Erlbaum Associates.

Martí, E. (2003). *Representar el mundo externamente. La adquisición infantil de los sistemas externos de representación.* Madrid, Spain: Antonio Machado.

Nussbaum, J., & Novick, S. (1982). Alternative frameworks, conceptual conflict and accommodation: Toward a principled teaching strategy. *Instructional Science, 11*, 183–200.

Olson, D. R. (1994). *The world on paper: The conceptual and cognitive implications of writing and reading.* Cambridge, UK: Cambridge University Press.

Peirce, C. S. (1931–1958). *Collected writings* (8 vols.). C. Hartshorne, P. Weiss, & A. W. Burks, (Eds.). Cambridge, MA: Harvard University Press.

Pérez Echeverría, M. P., & Scheuer, N. (2009). External representations as learning tools: An introduction. In C. Andersen, N. Scheuer, M. P. Pérez Echeverría, & E. Teubal (Eds.), *Representational systems and practices as learning tools* (pp. 1–16). Rotterdam, Netherlands: Sense Publishing.

Piaget, J. (1952). *The origins of intelligence in children* (M. Cook, Trans.). New York: International Universities Press.

Piaget, J. (2001). [1930]. *The child's conception of physical causality* (M. Gabian, Trans.). New Brunswick, NJ: Transaction Publishers.

Piaget, J., & Inhelder, B. (1969). *The psychology of the child* (H. Weaver, Trans.). New York: Basic Books, Inc.

Pozo, J. I., & Gómez Crespo, M. A. (1998). *Aprender y enseñar ciencia*. Madrid, Spain: Morata.

Rosebery, A., Warren, B., Ballenger, C., & Ogonowski, M. (2005). The generative potential of students' everyday knowledge in learning science. In T. Romberg, T. Carpenter, & F. Dremock (Eds.), *Understanding mathematics and science matters* (pp. 55–80). Mahwah, NJ: Lawrence Erlbaum Associates.

Russ, R. S., Scherr, R. E., Hammer, D., & Mikeska, J. (2008). Recognizing mechanistic reasoning in student scientific inquiry: A framework for discourse analysis developed from philosophy of science. *Science Education, 92*, 499–525.

Sarama, J., & Clements, D. H. (2009). *Early childhood mathematics education research: Learning trajectories for young children*. New York: Routledge.

Vygotsky, L. S. (1981). The genesis of higher mental functions. In J. V. Wertsch (Ed.), *The concept of activity in Soviet psychology* (pp. 144–188). Armonk, NY: M. E. Sharpe.

Wellman, H. (1990). *The child's theory of mind*. Cambrige, MA: MIT Press.

Wiser, M., Smith, C. L., Doubler, S., & Asbell-Clarke, J. (2009, June). *Learning progressions as tools for curriculum development: Lessons from the Inquiry Project*. Paper presented at the Learning Progressions in Science (LeaPS), Iowa City, IA.

Zhang, J., & Norman, D. A. (1994). Representations in distributed cognitive tasks. *Cognitive Science, 18*, 87–122.

The Ontology of Highlighting

David Hammer

Nemirovsky and Smith's chapter focuses on work by a mathematician, "Joseph," on 6j symbols. For me, the discussion recalled ideas I studied years ago about angular momenta, of "Clebsch-Gordon coefficients" and "irreducible representations." It's been a while, but I can still reconstruct some basic ideas: There are different ways to combine angular momenta mathematically, and depending on the physical system, some ways are more useful than others. The ideas I remember are about systems of two, such as an electron spin coupled with the angular momentum of its orbit in a hydrogen atom; 6j symbols are about systems of three.

For the electron or many other physical systems, the interaction makes it impossible to think of individual angular momenta as well defined individually. It's the system that's in a well-defined state, an "irreducible" new thing. That's what 6j symbols are about, forming that new thing, for a set of three angular momenta.

I read Nemirovsky and Smith's chapter first, out of the three in this section, and that's how I came to see them as describing "irreducible" states of cognition. So when Noble (this volume) writes that previous "findings challenge the assumption that students' representations simply reflect some form of internal representation," or when Nemirovsky and Smith write that they think of diagrams "as inseparable from enacted or implicit gesture, talk, drawing, body movement," I find myself thinking in terms of irreducibility.

It's a pretty tempting analogy. The idea that a mind or minds working with a representation form a new, inseparable whole seems like the combined irreducible state of the angular momenta in Joseph's explanation, in which it becomes impossible to attribute a fixed value to aspects of the individual angular momenta. At the same time, on both sides of this analogy, we're aware of the parts that make up the system—even if the state they're in all together is "inseparable." Those parts retain aspects of their individuality—there's still an electron in that atom, for example, and there's still Catherine and Melanie in Noble's interview.

One difference, clearly, is that the one can be made mathematically precise. 6j symbols are about specifying just how the three parts combine to produce the single whole. It's hard to imagine having that precision in our account of minds and representations.

Still, not to give up on the analogy, maybe we shouldn't think of a simple choice between adopting or rejecting a "dualistic" view of minds and representations. We can think about the situated dynamics of minds and representations as involving the rich complexity of interactions, and resist attributing stabilities we observe there to individual minds, without ruling out the involvement of individual minds in those dynamics—or the possibility of analytic separability. In their chapter, Gravel, Scheuer, and Brizuela analyze the events of reasoning in terms of interactions among various aspects of Iris's reasoning and her representations; they speak of the aspects as little "things" they are calling resources. Attributing those little things to Iris, they are adopting a kind of ontology, thinking that there exist little "pieces" (diSessa, 1988) of knowledge.

As it happens, I'm writing this just having watched video of "Energy Theater," in which Rachel Scherr and her colleagues (Scherr, Close, & McKagan, 2011) ask learners to enact "energy" with their bodies, its movement and transformations. The rules effectively require thinking of energy as conserved and having a location—the participants' bodies don't disappear! In this way, Energy Theater involves an ontology: Energy in this game exists as things that are like matter (historically, it came out of ideas about a substance scientists at the time called "caloric"). The video shows the learners struggling to keep track of how energy is conserved—if this much comes in, where does it go? Conservation is a natural affordance of the ontology. That is, if you think of energy as *stuff*, you're all set to think of it as conserved.

A great deal of the challenge to understanding angular momenta, and quantum mechanics more generally, is in the ontology: What kinds of entities and processes are in the model? Noble and Nemirovsky and Smith refer to how physicists think of light and electrons as both particles and as waves. It's worth clarifying, though, that these are new kinds of things. It's too simple to say that physicists sometimes think of light as a particle, when that seems useful, and sometimes as a wave. The theory coordinates those ways of thinking into a single, coherent whole: You can always think of light as a "particle-wave." It's just that in many situations you can tell easily that you don't need to, e.g., that thinking just in terms of waves will work.

I think I see an ontology implicit in Nemirovsky and Smith's analyses. When they write that "the use of Diagrams allows for indirect perceptuo-motor access to mathematical objects," they seem to be thinking of aspects of individuals interacting with aspects of inscriptions, as they documented in Joseph's work. To put this in the terms from Gravel et al., it sounds like they are saying that the Diagrams help Joseph tap into his perceptuo-motor resources and use them.

While I see an ontology in Nemirovsky and Smith's analyses, I'm confused by the ontology in Noble's account of highlighting and backgrounding: Where and in what form is the knowledge that is foregrounded or backgrounded? Students' external representations do not "simply reflect" internal representations, but the alternative might be that what students write and draw complexly reflects internal resources. That's how I understand the idea of amplification that Gravel et al. present, of the image of dots activating an aspect of Iris's reasoning, which has her using dots, and in this way a dynamic becomes stable. The stability occurs across Iris's mind and the materials, and Brian's questions to her in the situation of the interview. But we can still think of aspects of Iris's knowledge and reasoning as part of that stability. I think something similar was happening for me: I bring resources from my studies and experiences that are involved in my interactions with these chapters.

I enjoyed the moment in Nemirovsky and Smith's chapter when Smith asked Joseph why not add three spins all at the same time, and Joseph started his answer by saying, "We're sort of simple beings." It made me think, "We are simple beings trying to understand our complex minds!" Much as it helps to design representations, it also helps to design ontologies. Developing and refining a sense of "what kind of things are these" has been part of progress throughout the natural sciences, in understanding energy and quantum mechanics, and I believe it's likely to be essential for the learning sciences as well.

REFERENCES

diSessa, A. (1988). Knowledge in pieces. In G. Forman & P. Putall (Eds.), *Constructivism in the computer age* (pp. 49–70). Hillsdale, NJ: Lawrence Erlbaum Associates.

Scherr, R. E., Close, H. G., & McKagan, S. B. (2012). Intuitive ontologies for energy in physics. In C. Singh, M. Sabella, & P. Englehardt (Eds.), *AIP Conf. Proc. 1413 Physics Education Research Conference* (pp. 343–346). Omaha, NE: American Institute of Physics.

Part IV

REPRESENTATIONS AS SCAFFOLDS AND SUPPORTS

CHAPTER 10

Children's Design Constructions as Representations of Science Ideas

Kristen Wendell

Imagine a 3rd-grade student holding a plastic rectangular frame that he and a partner have assembled out of construction kit toys, studded with pegs, and spanned with rubber bands. He plucks the rubber bands and listens for the pitch of their sounds:

> *Devin*: How is this, this is, this [*plucking a thin band stretched partially across the frame*] is higher than this [*plucking another thin band stretched across the entire frame*] somehow!
>
> *Kristen*: Which one?
>
> *Devin*: [*Plucking the thin, partially stretched band*] Than this [*plucking thin, fully stretched band*].
>
> *Kristen*: This (partially stretched) one is higher? Which, what's different? You're having a hard time figuring out why?
>
> *Devin*: [*Suddenly turns around and looks at a large poster with the class's notes about pitch, and moves toward it to read.*] Oh, now I know. This (thin, partially stretched) one's higher because—that is weird! Kind of weird.
>
> *Kristen*: Devin, what do you mean? Devin, what did you figure out?
>
> *Devin*: The smaller it is, the higher it is. So this is our high (thin, partially stretched band), this is our medium (thick, partially stretched band), and this is our low (thick, fully stretched band).

In the excerpt above, Devin is talking to me while he is in the midst of completing an engineering design challenge. He has been working with his partner, Jamie, to create a musical instrument that can play at least three different pitches—"low," "medium," and "high." Devin and Jamie intend for their instrument, which is inspired by a real-life guitar demonstrated by their teacher, to produce different pitches via rubber bands set to varying tensions. When Devin tests out his and Jamie's instrument, he expects the band stretched across only part of the frame to sound lower-pitched than the band stretched across the entire frame. This is because he thinks (rightly so) that partial stretching puts less tension in the band than does full stretching. However, Devin is not considering that the length to which a rubber band is stretched also affects its pitch. The variables of tension and length are interacting with each other, and in the current configuration, length has the dominant effect. The partially stretched rubber band sounds higher-pitched because it is shorter, but Devin does not realize this at first, and he is quite surprised by how his instrument works. His surprise sends him running, almost literally, to his classroom's collective record of findings about sound. There he finds the reason for the discrepancy between his expectations and the instrument's performance: Although lesser tension causes lower pitch, so does greater length. Devin now makes sense of this "weird" relationship between size and pitch: The smaller it is, the higher it is. Through Devin and Jamie's musical instrument construction, new understandings of physical phenomena are emerging. My goal in this chapter is to argue that children's design constructions can be representations of these understandings. Based on my work engaging children in engineering design, in this chapter I will suggest that the tangible result of a child's engineering process can be viewed as an external indicator of his or her ideas. Furthermore, those ideas are constructed along with, and transformed by, the three-dimensional representation.

DESIGN CONSTRUCTIONS DEFINED

Devin and Jamie's musical instrument is an example of what I will refer to in this chapter as children's *design constructions*. Design constructions are tangible, three-dimensional artifacts that result from some kind of engineering design process and that are created by children to perform specific functions or solve specific problems (Dym, 1994; Roth, 1996). They typically take the form of rough prototypes tested against the requirements of the design problem (Benenson, 2001) and accompanied by writing, drawing, or speaking that would enable prototype replication (Dym, 1994). A child's prototype may be the actual solution to a design problem, such as a musical instrument used to play an actual song or a plastic bottle terrarium that will actually be used for a life sciences study. Or it may be only a functional model of a solution, such as a LEGO® car that can climb a ramp to model a real car with the ability to climb hills. Children's design constructions may also be called *design artifacts*; I use the two terms interchangeably in this chapter.

The creation of a working engineering design product requires the development and use of conceptions about physical phenomena; therefore, a design construction can indicate the range of such conceptions held by the designer. This claim implies that while the process of designing and constructing 3-D artifacts is an engaging learning experience, the completed constructions themselves are also important to learning and thinking. Thus, I view design constructions as cognitively useful representations. This view is consistent with the perspective discussed elsewhere in this book that any representation produced by an individual serves some purpose for that individual's cognition.

ROLES ASSIGNED TO DESIGN CONSTRUCTIONS

Let me take a step back here and briefly review how others have described the role of design artifacts in students' science learning. In Penner and colleagues' (1997, 1998) design-based modeling approach to science instruction, students are challenged to create working replicas of the object being investigated by the class, such as the human elbow. Penner and his colleagues view these student-built models as mediums for the study of mechanism. This means that the design constructions further the students' investigation of physical cause and effect.

According to Roth (1996), design constructions do more than just enable the study of mechanism. They can become thinking tools, indicators of process steps, and platforms for classroom discussion and sense-making. Roth (1996) writes, "Emerging artifacts constitute a focus and backdrop for students' discursive activities of talking, pointing, and gesturing, that allow them to make sense of each other's utterances and to negotiate shared meanings in the face of ambiguity" (p. 157).

In their work on middle school engineering competitions, Sadler, Coyle, and Schwartz (2000) see the design construction as a test bed for students' science conceptions. If students' science conceptions are non-normative, their devices may not work. Sadler and colleagues (2000) contend, "Design is a form of cognitive modeling that crystallizes a conceptual model into a physical embodiment, either on paper or as a physical entity" (p. 304).

Finally, in work on project-based science (Krajcik & Blumenfeld, 2006) and Learning by Design™ (Kolodner, Camp, Crismond, Fasse, Gray, Holbrook, Puntembakar, & Ryan, 2003), design constructions serve in large part as motivators. The challenge of creating a functioning artifact provides incentive and opportunity for scientific reasoning and learning. That is, design constructions motivate students' engagement in science, and the manipulation of the constructions leads to the manipulation of science ideas. Ideally, this leads to deeper understanding.

Looking beyond the science education literature, work on distributed cognition also speaks to the role that design constructions may play in the instructional setting. Distributed cognition theory offers the helpful construct of *cognitive residue* (Bell & Winn, 2000). Humans generally assume that when they appropriately use tools (such as pencils, paper, calculators, or books) as part of a cognitive act, they are more productive and their intellectual capabilities are enhanced. However, what is not quite as obvious is that those tools can leave a sort of residue that supports intellectual activity later, even when the tools are no longer present (Bell & Winn, 2000).

Distributed cognition theorists often focus on computers as technological tools, but many other human-created artifacts, including children's design constructions, can offer a source of cognitive residue. Although design constructions serve as an end in themselves and are not necessarily intended to be cognitive tools, they may provide a lasting cognitive effect for students who develop new understandings to solve a design problem. It is reasonable to consider that a design construction embodies a student's emerging conception, and at a later date, recalling the artifact triggers the conception as well. For example, imagine a 3rd-grader who builds a miniature model house that is thermally insulated (i.e., its internal temperature does not drop excessively when it is surrounded by ice cubes). By the time she formally studies the topic of heat transfer (probably in high school), she no longer has her miniature house from 3rd grade. However, her experience interacting with the thermal properties of that house has left a cognitive residue that amplifies her ability (Cole & Griffin, 1980) to make sense of complex phenomena such as conduction and convection.

Thus far, we have seen that other researchers view design constructions alternately as sources of cognitive residue, motivators of learning, test beds for idea accuracy and usefulness, records of students' steps, or springboards for discourse about physical mechanisms. My main claim in this chapter will be that while artifacts constructed by students can be construed to serve all these purposes, they also fundamentally exist as representations of students' understandings of physical phenomena. As a result, design constructions serve not only to reflect these understandings but also to amplify and transform them.

The sociocultural view of learning supports this notion that design constructions represent students' ideas in science. Vygotsky (1962) proposed that the use of "cultural tools" alters individuals' mental functions. Cultural tools include invented hardware such as spoons and hammers, symbol systems such as speech and writing, and patterns of social interaction such as the rules of conversation and the management of resources. The use of cultural tools mediates humans' interactions with the physical and social environment, and thus mediates cognition. In the activity of engineering design, several cultural tools are present, including the design constructions themselves as well as the conversation and drawing (and other external representations) used to plan and describe the constructions. The design constructions support students in *pursuing* novel and interesting interactions with the tangible, material world. Conversation and other symbolic representations of the design constructions support children in *making sense of* these novel interactions. Thus, engineering design offers tools that facilitate students' efforts to construct scientific understandings.

For instance, consider an example from my research on children's musical instrument engineering. Through interviews on how musical sounds are produced,

I have found that many children's initial exposure to scientific discourse about sound occurs when an adult introduces the concept of *vibration* and announces that all sound is made from vibrations. As a result, quite a few children readily use the term *vibrations* to answer my questions about how sounds are created. However, I observe that they cannot explain what *vibration* means, nor can they identify what object is vibrating. The word has preceded an understanding of the underlying meaning. Fortunately, this state of affairs changes after the children have an opportunity to construct musical instruments out of small building toys, rubber bands, and balloons. The children interact with these hardware tools, and they use other cultural tools to write, draw, and speak about them. This tool use (in the Vygotskian sense) facilitates the young engineers' discovery that to cause their percussion and string instruments to make sounds, they must impart on them some sort of back-and-forth motion, perhaps by plucking a rubber band or balloon membrane and letting it reverberate, or by shaking a handmade maraca back and forth. After these experiences, when the concept of *vibration* is mentioned again, the children are able to use it productively to predict what other objects will be good noisemakers. In a reciprocal way, the word "*vibration*" is a tool for targeting attention to the back-and-forth motion of the design constructions, and the constructions are tools for making meaning of this technical term.

CHILDREN'S MUSICAL INSTRUMENT ENGINEERING

Now I draw upon further evidence from my research to illustrate how design constructions can function as representations of science understanding. The evidence comes from a study of 3rd- and 4th-grade students' work during a science curriculum unit called *Design a Musical Instrument: The Science of Sound*. This unit was developed as part of a larger research program on the relationship between science learning and engineering design activity in the upper elementary grades (Wendell et al., 2010). It has provided a rich context for exploring children's design constructions. In the opening lesson of the unit, students learn that their engineering design challenge is to create a new musical instrument that can play at least three different notes and contribute to a classroom band. Over the next six lessons, students conduct a series of guided, design-based investigations to explore how sounds are produced, transmitted, and varied across different sound producers. Using LEGO® construction kit elements and craft materials, they build a miniature drum, pan pipe, rubber-band guitar, and maraca. They explore the structural design of these instruments, observe how they look and sound when played, and identify the characteristics of the sounds they make. With the intention of fostering science conversations among students, the curriculum encourages students to explain how physical characteristics are connected to sound characteristics. For example, how does the size of an object influence the pitch of its sound?

Throughout the unit, students are encouraged to consider how these relationships between visible and audible characteristics can inform their design of a new musical instrument. In the unit's two concluding lessons, students design, construct, and demonstrate musical instruments of their own invention. The unit is intended for 3rd-grade students and requires approximately 12 hours of classroom time for its nine lessons.

To illustrate how the students' musical instrument design constructions represent and interact with understandings of physical phenomena, I will draw from the work of two student pairs—Devin and Jamie, mentioned earlier, and Casey and Morgan. The students in both pairs attend the same K–8 public school in an urban district in the northeastern United States. In the school community, approximately 80% of students are eligible for free or reduced-price lunch, and about 60% learned English as a second language. The two pairs featured in this chapter have different teachers, but the teachers, Mrs. N and Mrs. H, are close colleagues with adjoining classrooms. They jointly attended the professional development workshop on the *Design a Musical Instrument* unit. Each student pair worked together for all of the lessons in the *Design a Musical Instrument* unit, including the final design challenge highlighted in this chapter. The key activities of the final design challenge lessons are listed in Table 10.1.

I will begin with Devin and Jamie. After presenting their completed musical instrument, I will discuss the ways in which their earlier iterations showcase their evolving ideas about the science of sound. Then I will explain how Casey and Morgan's final musical instrument represents a very different understanding of sound from that developed by Devin and Jamie.

RUBBER BANDS AND PITCH: DEVIN AND JAMIE'S INSTRUMENT

Devin and his partner Jamie created a design construction that would be classified as a stringed instrument. Their final construction, shown in Figure 10.1, consisted of eight LEGO® beams stacked together in a platform, four pairs of connector pegs spanning the length of the platform, and four rubber bands of different widths and tensions suspended from those connector pegs. Devin and Jamie relied on only one material—rubber bands—to meet the design requirement of at least three distinct pitches. In their oral presentation, they identified accurately four distinct pitches. They also explained accurately that the band that produced the highest pitch did so because it was thinnest and had the most tension. At one point in their explanation, Devin also introduced the idea that the lowest-pitch band produced its pitch because it was short. However, after Mrs. N said this made her "confused," he revised his explanation to say that this band produced a low pitch because it had only a small amount of tension.

Table 10.1. Key Activities of the Final Design Challenge in the *Design a Musical Instrument* Unit

Day	Activities for the Day
1	With partner, discuss ideas for solving the musical instrument design challenge Write about and draw a plan Have plan approved by teacher
2	Build first version of musical instrument Test instrument Improve instrument
3	Continue to improve instrument If finished, create a poster that shows instrument and labels its pitches
4	Finish instrument poster Practice oral presentation for class design expo
5	Describe and demonstrate instrument in the design expo

Figure 10.1. The Final Musical Instrument Created by Devin and Jamie, with Pitches Labeled as They Were Identified by Devin During Their Oral Presentation

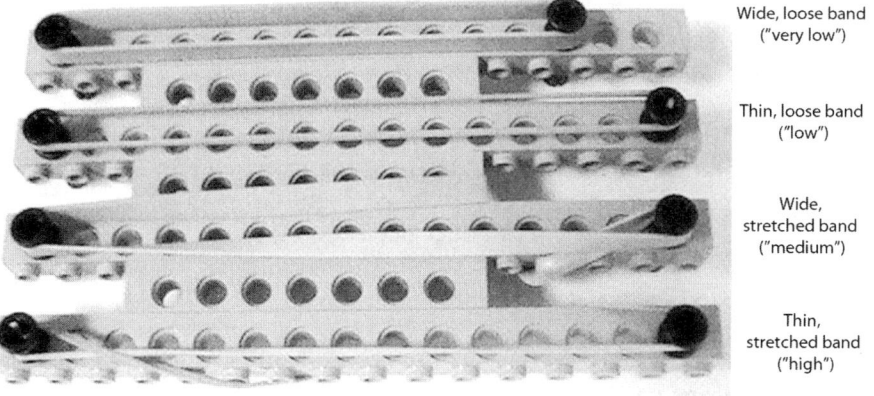

Wide, loose band ("very low")

Thin, loose band ("low")

Wide, stretched band ("medium")

Thin, stretched band ("high")

Day 1 for Devin and Jamie

Prior to presenting their final instrument, Devin and Jamie created several intermediate design constructions, shown in Figure 10.2. Each iteration embodies aspects of the students' current understanding of how instruments produce sounds with different pitches. On Day 1, their idea is that they need a square frame and several different sound-producing elements to produce several different pitches. No photo was taken of their Day 1 construction, but Figure 10.2 shows the pair's drawing of it. In the conversation that accompanies their creation of this drawing, Devin mentions length and thickness, but together Devin and Jamie do not come to explicit consensus on how their sound-producing elements will be varied. However, there seems to be a developing plan. Devin suggests that they will pick two thick and two thin rubber bands, and by moving them "however we want," they will achieve "high, very high, and medium high, and medium low" pitches.

Jamie: Um, we're gonna make a box—

Kristen: Yeah?

Jamie: And it's gonna have connector pegs. And, I think we're gonna have elastics? [*Looks at Devin.*]

Devin: Yeah, elastics.

Kristen: You're gonna have elastics?

Devin: [*Stands up and looks at Jamie's page.*] Just draw—we're not gonna put the elastics on. Wait, we can't, we don't know how to draw the elastics, 'cuz they're gonna, we'll just move them, however we want—

Jamie: I know how to draw them! Like this! [*Draws an elongated oval across the rectangle he has drawn on his page.*]

Kristen: Okay. Does that work, Devin? See how Jamie just drew it?

Devin: Yeah. Yeah.

Kristen: What did you, what do you mean you're gonna move them?

Devin: Yeah, like it can be here [*points to oval on page*] and we move it to there [*points to another place on rectangle*] and stuff like that.

Kristen: To make, why will you move it?

Devin: To make different notes. We're gonna have two thin ones and two long—and two thick ones.

Kristen: Two thin ones and two thick ones?

Devin: Yeah. It's a guitar.

Kristen: How did you decide on that?

Devin: [*Looks at Jamie's page. Jamie and Devin look at each other, and both shrug.*] Uh, it'll make it easier, 'cuz we can make it high, very high, and medium high, and medium low.

Figure 10.2. Sequence of Designs Prior to Devin and Jamie's Final Musical Instrument Construction

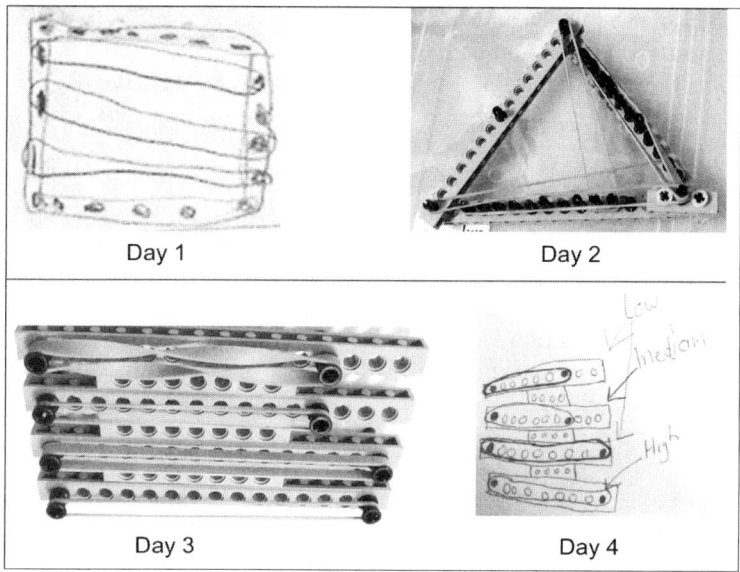

Day 2: A Triangle

Over the course of Day 2, Devin and Jamie make a few adjustments. They spend the first 26 minutes of Day 2 working on a square frame for their instrument. However, it keeps collapsing at its corners, and despite repeated attempts, Devin and Jamie cannot find a secure way to attach diagonal bracing to it. Devin cries in exasperation, "Our thing's not strong enough!" and Jamie responds by repeating a suggestion that Devin had dismissed on Day 1: "I knew we should have done it out of a triangle!" Devin accepts the suggestion this time, and within 5 minutes, he converts the fragile square frame into a triangular frame. This design construction represents one key change in the pair's understanding of the science of sound: Musical instruments must have stable frames, square frames are quite difficult to stabilize, and triangles are inherently stable.

The way Devin and Jamie add rubber bands to their triangular construction also suggests that their understanding is evolving to include an explicit set of physical variables—length and thickness—that affect a sound-producing element's pitch. They stretch four rubber bands roughly along the legs of the triangle. As planned by Devin on Day 1, two of the bands are thin and two are thick, and the students take pains to stretch them to slightly different lengths. In the following excerpt, Jamie orally articulates a relationship between length and pitch, and Devin attempts to make sense of it by plucking the corresponding bands on their construction. Susan is a researcher who has been checking in with them periodically.

> *Jamie*: Low should be long, uh, high should be short, and I think that's all. And medium should be medium.
> *Devin*: Low [*plucking thick, longest band*]. Medium [*plucking thinner band*]. Low [*plucking another thick band*]. [*Stretches another thin band along leg of triangle.*]
> *Susan*: So which one's which?
> *Devin*: Look. Medium (thick band stretched across interior of triangle). Low (thick band stretched along entire leg of triangle). And high (thin band also stretched along entire triangle leg).
> *Susan*: So why is that (thin stretched) one high?
> *Devin*: Because it's thin and it's long.
> *Susan*: And what about the others?
> *Devin*: It's (thick band stretched across interior) thick and it's long, so it's medium.

In this exchange, Devin appears to be adapting Jamie's oral remarks about length to his own idea that thickness determines pitch. As can be seen in the photograph of Day 2's triangular construction, the variation in rubber-band length is quite subtle, and it is not accomplished in an orderly or systematic fashion. Further, the variation in width (i.e., thickness) of the rubber bands is interfering with the variation in length. Although Jamie articulates a simple, direct relationship between *length* and pitch, Devin's main idea is that *thickness* determines pitch. Consequently, Devin is inconsistent in identifying which of the actual bands has the highest pitch. The pair's design construction reveals their progressing but still "muddy" collective understanding of how physical dimensions affect pitch. They now know that either thicker *width* or longer *length* can cause lower pitch. But what if a band's length is extended at the same time that its width is reduced? There is not yet evidence that Jamie and Devin have considered the variable of *tension*, which can be the dominant factor in determining pitch.

Day 3: A Platform

On Day 3, the students' reasoning about pitch shifts, and this shift is both aided by and represented in their next design iteration. They abandon the triangular frame because Devin finds it too difficult to keep track of which rubber band has which pitch when they are "crisscrossed" at acute angles. (Devin is attached to the idea of stretching the rubber bands along the edges of their instrument's frame; he cannot make sense of Jamie's idea that the triangle would provide for a nearly infinite number of lengths if the rubber bands were stretched across its interior.) As shown in Figure 10.2, on Day 3 they create a new rectangular frame that allows Devin to more easily keep track of the length of the rubber bands. This new construction affords experimentation with pitch in a more systematic manner. For Devin and Jamie, this is a more scientific strategy for exploring their ideas. As they tweak and adjust their instrument design, they can likewise tweak and adjust their understandings of how different sounds are generated. Their design constructions mediate their thinking about sound. The following excerpt contains their conversation just before building this rectangular frame, which Devin calls a "platform." Quite literally, this platform for rubber bands is also a platform for the students' experimentation.

Jamie: Devin wants to do it all over.

Devin: Because this [*holding up triangle frame*] ain't working exactly, 'cuz it's all mixed up.

Kristen: What is it, what do you mean, it's all messed up?

Devin: It's mixed up. It's hard to play, 'cuz every [*gesturing as if stretching bands across the triangle*], they're over here [*gesturing along one leg of triangle*], they're over here [*gesturing along other leg of triangle*], they're over here [*gesturing across triangle, from leg to leg*].

Kristen: Oh, they were like crisscrossed?

Devin: Yeah, 'cuz I was crisscrossing them.

Kristen: Could you, do you mean you're gonna build a whole new triangle, or just put the rubber bands in different places? Could you talk about it first?

Devin: [*Turns to Jamie.*] Make something like a straight platform, instead of like a box or something. Like this [*holding two beams in a straight line*], straight, a big platform.

Kristen: Show me (along the beams) where your rubber bands are gonna go. What do you mean?

Devin: I don't have that yet. We both don't know where it's gonna go because we haven't *built* yet [emphasis in the original].

Devin's last comment is intriguing. It tells us something about Devin's own perspective on design constructions as representations of ideas. He appears to be claiming that he cannot "have" an idea about how rubber bands will vary in pitch until he is physically manipulating them. Devin's perspective is that not only does his design construction *represent* his idea, but also that his idea *comes into existence* at the same time as the design construction. This view is reminiscent of Nemirovsky's (2009) discussion of the embodiment of ideas in representations; in this way, Devin's ideas and representations are inseparable.

Later on Day 3, Devin and Jamie behold their completed platform and debate the relative pitches of its four rubber bands.

Jamie: This one [*points to band on instrument, then starts plucking thin and thick bands that are partially stretched*]. These two are the same, I think.

Devin: See, this one (thin, partially stretched band) is the highest, and this one (thick, fully stretched band) is the lowest.

Jamie: [*Plucks all four bands.*] Actually, this [*points to thin, fully stretched band*] is the highest because when you put tension, it's higher, actually.

The above conversation suggests that interestingly, the instrument now represents different ideas for Devin and Jamie. On the one hand, it embodies Jamie's understanding that the variable of tension can dominate the variable of length, and that greater tension can cause higher pitch even when the tension results in a longer sound-producing element, whose length would otherwise relate to lower pitch. On the other hand, it embodies Devin's (partially correct) thinking that size is the dominant variable in determining musical pitch: The thinner and shorter something is, the higher its pitch.

Final Adjustments

On Day 4, Devin and Jamie face the task of creating a poster with a drawing of their final musical instrument and labels of its pitches. This assignment leads to more conversation between the two of them about which band produces which pitch, and it pushes Devin's understanding closer to Jamie's. Although he does not *orally* express the tension–pitch relationship with great consistency, Devin does *physically* make changes to their design construction that reveal his recognition of the role tension plays in determining pitch. As shown in Figure 10.2, he wraps the two bottom rubber bands around the back of the platform, which increases their tension even more and thus makes their pitches distinctively higher than the two upper bands. In the following excerpt, he describes this design feature to his teacher, Mrs. N.

Mrs. N: Okay, uh, tell me about your instrument. Was that always your original idea? Is that what you were thinking you were gonna do to start with?

Jamie and Devin: [*Shake heads no.*]

Mrs. N: Where did you start? What was your first idea?

Devin: We had, uh, it was a triangle, but it was too mixed up. And—

Mrs. N: It was too mixed up. What do you mean?

Jamie: [*Nudges Devin.*] Actually we built a square. But, we nev—

Devin: It just kept on folding.

Jamie: Yeah, because whenever we stretched the elastic, and put it one end to the other, it keeps on going like a diamond [holding two hands out and then moving them down and in, as if breaking a stick].

Mrs. N: So it wasn't a sturdy frame—

Devin: It wouldn't go to a diamond, it would more like go to st-, flat.

Mrs. N: It kept folding up, yeah, so not holding together. So then what was the second idea?

Devin: A triangle. But it was too mixed up. Too mixed up.

Mrs. N: What do you mean by too mixed up?

Jamie: . . . Because a lot of the elastics were like going everywhere.

Mrs. N: And did they make different notes, or you couldn't tell?

Devin: They were making different notes, but it was hard to play, 'cuz we tried to play high, medium, low, and we went high, low, medium.

Mrs. N: So it was too confusing? Then, was this your third idea? . . . What were you thinking?

Devin: We were looking at, uh, Ethan, and we saw him bringing a big, uh, wooden thing to attach (to Ethan's LEGO® creation) so we took his other idea [Ethan's instrument included a LEGO® platform with rubber bands, but next to that he attached a wooden structure so the bands could be stretched even farther].

Mrs. N: So just to use the neck [i.e., the LEGO® platform part of Ethan's instrument], and stretch it out?

Devin: [*Nods.*]

Mrs. N: Um, how could you make your higher note even higher?

Devin: Because, if it was like this [*stretched only along front of platform*], if it wasn't like this [*stretched around to back of platform*], then we'd go around [*pulling band to back side of platform*].

Mrs. N: So you added more tension to it?

Devin: [*Nods.*]

Although Devin makes sense of and agrees with his teacher's use of the word "*tension,*" Devin's own oral expression does not include the term. We must examine his design construction to find a representation of tension as an emerging part of his conceptual framework for pitch. As Devin and Jamie demonstrate their final instrument construction to their classmates and teacher, the construction serves as a repository of the understandings they developed. The above conversation illustrates how their final instrument serves as a record of their changing design ideas and scientific understandings over the course of the engineering challenge. Their instrument also reveals their newly complex conception of pitch, in which length, thickness, and tension can all play a role. By contrast, the final instrument created by Morgan and Casey suggests a conception of pitch in which these physical dimensions play no role. Instead, Morgan and Casey's instrument represents their idea that material kind (i.e., the substance from which an object is made) is the primary determinant of pitch.

THREE MATERIALS, THREE SOUNDS: MORGAN AND CASEY'S INSTRUMENT

Morgan and her partner Casey constructed a musical instrument, shown in Figure 10.3, that had features of both the percussion and string families. They created a triangular frame with three long beams, stretched a balloon membrane across both sides of the frame, placed loose pegs in between the two membranes, and wrapped a rubber band around the exterior of the frame from one leg of the triangle to another. They played their instrument through a combination of shaking (the loose pegs), plucking (the rubber band), and striking (the balloon). Essentially, Morgan and Casey relied on three entirely different materials in their attempt to meet the design requirement for three distinct pitches. But when they played their instrument, the shaking sounded high-pitched, but the plucking and striking sounds were nearly indistinguishable by pitch. Thus, the instrument did not fully meet the design requirement. Further, Morgan and Casey misidentified the rubber-band pluck as the instrument's highest pitch, and they did not attempt to explain the physical cause behind that claim.

Unlike Devin and Jamie, who iterated through several prototypes, Morgan and Casey produced only one design construction over the course of working on the final musical instrument engineering challenge. Initially thinking that their instrument had to be a miniature replica of a real instrument, they spent a long period of time discussing what kind of instrument to build (e.g., whether to build a "guitar," a "drum," or a "xylophone"). When they did realize that their instrument could be a design from their own imaginations, they exchanged just a few sentences about how it would produce its sound. And as suggested by this excerpt, which includes a question posed to a researcher named Mary, these sentences did not reveal consideration of the mechanism behind pitch.

> *Casey*: Can the elastic (rubber band), like, can you like, put like the elastic on like, like, something like that [stretches rubber band open with two hands and moves it toward the triangle frame], and then, like, put the balloon on the elastic, and then put something to hold it?
> *Mary*: Maybe.
> *Casey*: Like, thi-, and the bal-, the elastic's like that [using both hands to stretch it wide open], and there's something holding the elastic, and then we put the balloon in.
> *Morgan*: This can hold it up [holds a peg upright as if to show it sticking out of something].

During the class sessions spent building their final instrument, Morgan and Casey never discuss how the physical characteristics (e.g., size, tension) of their instrument's components would affect its pitch. Rather, after adding its

Figure 10.3. The Final Musical Instrument Created by Morgan and Casey

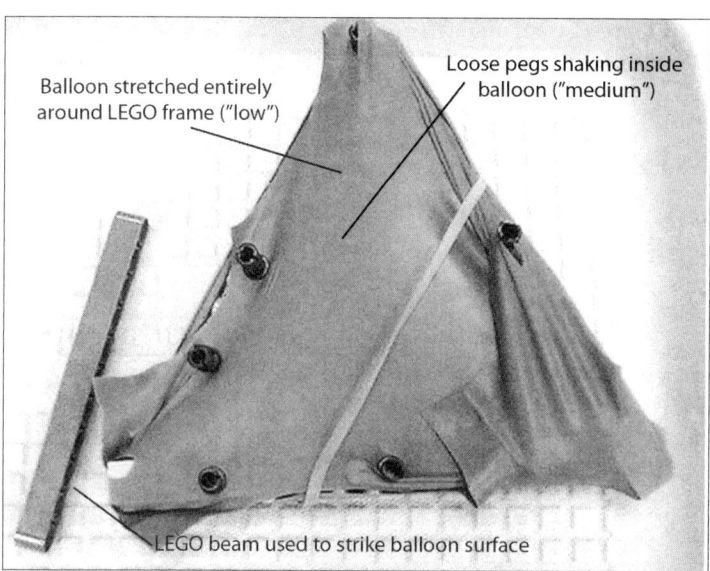

sound-producing elements (the pegs, balloon, and rubber band), they listen to its sounds and identify one as the high pitch, one as the medium pitch, and one as the low pitch. When they demonstrate their instrument to their classmates on the last day of the unit, they point out the correspondence between the three different components of the instrument and its three different pitches.

> *Casey*: The rubber band is the high sound [*points to the band stretched across the frame*].
>
> *Mrs. H*: Let's listen.
>
> *Casey*: [*Plucks the rubber band.*] And when you shake it it makes the medium [*makes the pegs rattle around inside the instrument*].
>
> *Classmate 1*: Cool! Oh, that's cool! You put the pegs inside the balloon. Cool.
>
> *Casey*: And when you go like that [*taps on the stretched balloon*], it's the low.
>
> *Mrs. H*: Okay, Morgan, will you play them? High, medium, low.
>
> *Morgan*: [*Plucks rubber band, shakes instrument, and taps balloon in sequence with the teacher's words.*]
>
> *Classmate 2*: Why did you put the (pegs) inside the balloon instead of like, like, um, outside?
>
> *Mrs. H*: Ah, I actually like that question. How come you decided to make the maraca part inside the drum part?
>
> *Casey*: So we could, so, the drum part and the maraca could make noise together.

Each element of the instrument is made of a different material, and Morgan and Casey attribute a different pitch to each element. Morgan and Casey's design construction represents an understanding of pitch as determined largely by material kind.

DESIGN CONSTRUCTIONS AS WINDOWS AND TOOLS

There are, of course, several limitations to viewing children's design constructions as representations of their understandings of physical phenomena, and I want to mention a few of the most important. First, we often need to consult additional modes of representation to interpret the ideas embodied in a design construction. For example, to make full sense of the musical instruments presented in this chapter, I also analyzed the students' oral discourse, and I looked at the written text they had included on their design posters, shown in Figures 10.4 and 10.5. Second, I have mostly described constructions as representing the understanding of a *pair* of students. In reality, each individual student has his or her own distinct conception of pitch. When a construction is the joint output of multiple children, it is difficult to disentangle one child's understanding from another's. In addition, we saw that Devin and Jamie also borrowed an idea from Ethan, a student outside their dyad, so in fact, the boundaries between individuals' thinking are quite hard to establish. Finally, a limitation shared by any mode of external representation is that one production seldom represents the totality of an individual's understanding. There are quite likely other aspects of these children's conceptions of sound that they did not choose to represent in their design constructions or in their related conversation and writing.

Certainly, Morgan and Casey, as well as Devin and Jamie, have many more ideas about the science of sound than are represented in their miniature musical instruments or discussed in this chapter. I would not suggest in any way that design constructions be used as the only access point to conceptions. However, as suggested prominently by Kolodner and colleagues (2003) and by several other researchers (e.g., Roth, 1996; Sadler et al., 2000), when children have an operational understanding of how a physical variable relates to the functionality of a device, their design constructions likely reveal that understanding. The assumptions and decisions embodied in their designs are a window into their understandings of the science. We have seen in this chapter that Devin and Jamie could express how tension and size affect musical pitch, and their musical instrument construction made use of variations in tension and size. We have also seen how Morgan and Casey did not express a relationship between pitch and these physical variables, and their musical instrument construction relied only on differences in material kind. As these examples show, children's design constructions—both in final form and in intermediate iteration—reveal a great deal about how children understand scientific

Figure 10.4. The Poster Created by Devin and Jamie to Document Their Final Musical Instrument Design. Note That the Pitch Labels Are Slightly Different from the Oral Presentation.

Figure 10.5. The Posters Created by Morgan and Casey to Document Their Final Musical Instrument Design

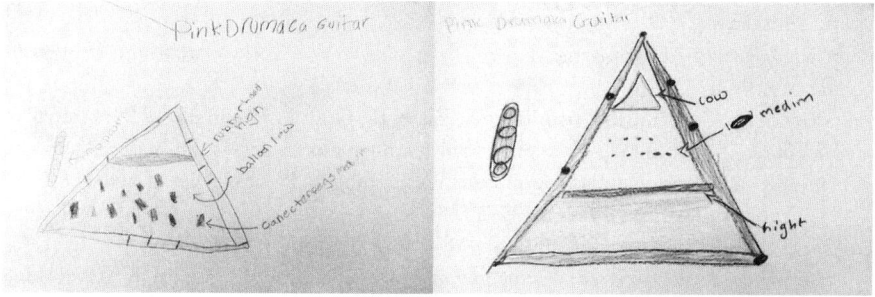

phenomena. Design constructions can do more than this, though. When treated dynamically by their designers, they can also serve as tools for transforming thinking. For example, in the first case in this chapter, Devin and Jamie used their Day 3 design—a straightforward platform—to systematically explore the relationships among length, thickness, tension, and pitch. By contrast, in the second case, Morgan and Casey treated their instrument design as a static reflection of their ideas, and as a result they lacked an important tool for shifting their reasoning about the science of sound. During children's engineering design, a complex set of interactions—among students, their physical materials, their ideas, and the intervention (or lack thereof) of adults—greatly influences the degree to which the design construction can transform students' thinking. However, when these interactions occur productively, students' design constructions not only reveal their understandings of physical phenomena, but they also advance those understandings.

ACKNOWLEDGMENTS

This material is based upon work supported by the National Science Foundation under Grant No. 0633952. The opinions, findings, and recommendations expressed in this material are those of the author and do not necessarily reflect the views of the National Science Foundation.

REFERENCES

Bell, P., & Winn, W. (2000). Distributed cognitions, by nature and by design. In D. Jonassen & S. M. Land (Eds.), *Theoretical foundations of learning environments* (pp. 123–142). Mahwah, NJ: Lawrence Erlbaum Associates.

Benenson, G. (2001). The unrealized potential of everyday technology as a context for learning. *Journal of Research in Science Teaching, 38*(7), 730–745.

Cole, M., & Griffin, P. (1980). Cultural amplifiers reconsidered. In D. R. Olson (Ed.), *The social foundations of language and thought.* (pp. 343–364). New York: W.W. Norton & Company.

Dym, C. L. (1994). *Engineering: A synthesis of views.* New York: Cambridge University Press.

Kolodner, J. L., Camp, P. J., Crismond, D., Fasse, B., Gray, J., Holbrook, J., Puntembakar, S., & Ryan, M. (2003). Problem-based learning meets case-based reasoning in the middle-school science classroom: Putting Learning by Design™ into practice. *Journal of the Learning Sciences, 12*(4), 495–547.

Krajcik, J. S., & Blumenfeld, P. C. (2006). Project-based learning. In K. L. Sawyer (Ed.), *The Cambridge handbook of the learning sciences* (pp. 317–333). Cambridge, UK: Cambridge University Press.

Nemirovsky, R. (2009). Remarks on external representations as learning tools. In C. Andersen, N. Scheuer, M. P. Pérez Echeverría, & E. Teubal (Eds.), *Representational systems and practices as learning tools* (pp. 281–296). Rotterdam, Netherlands: Sense Publishing.

Penner, D., Giles, N. D., Lehrer, R., & Schauble, L. (1997). Building functional models: Designing an elbow. *Journal of Research in Science Teaching, 34*(2), 125–143.

Penner, D., Lehrer, R., & Schauble, L. (1998). From physical models to biomechanics: A design-based modeling approach. *Journal of the Learning Sciences, 7*(3/4), 429–449.

Roth, W.-M. (1996). Art and artifact of children's designing: A situated cognition perspective. *Journal of the Learning Sciences, 5*(2), 129–166.

Sadler, P. M., Coyle, H. P., & Schwartz, M. (2000). Engineering competitions in the middle school classroom: Key elements in developing effective design challenges. *Journal of the Learning Sciences, 9*(3), 299–327.

Vygotsky, L. S. (1962). The development of scientific concepts in childhood. In L. S. Vygotsky, *Thought and language* (pp. 82–118). Cambridge, MA: MIT Press.

Wendell, K. B., Connolly, K. G., Wright, C. G., Jarvin, L., Rogers, C., Barnett, M., & Marulcu, I. (2010). Incorporating engineering design into elementary school science curricula. *2010 Proceedings of the American Society for Engineering Education Annual Conference & Exposition.* Louisville, KY: American Society for Engineering Education.

CHAPTER 11

Exploring 1st-Grade Students' Drawing and Artifact Construction in the Engineering Design Process

Merredith Portsmore

Efforts to bring engineering into Kindergarten through grade 12 (K–12) education include the creation of standards at the state and national level (International Technology Education Association, 2002; Massachusetts Department of Education, 2001) as well as the development of activities and curriculum that introduce students to the engineering design process—a model of the practices that adult professional engineers engage in when designing a product or process that addresses an identified problem or need. Engineers express their design ideas at formative and summative stages of the engineering design process through the creation of representations (e.g., drawings, Computer Assisted Design [CAD] models). Efforts to introduce children to engineering design typically include asking children to represent plans for their preliminary ideas through drawing. However, current research is divided as to whether children, particularly early elementary students, can engage in meaningful planning via drawing for engineering design problems.

This chapter focuses on a portion of an exploratory study that looked specifically at the drawings 1st-grade students created when presented with a hands-on engineering design problem prior to constructing their solution. The goal of the study was to investigate if the activity 1st-grade students engage in prior to constructing an artifact indicates that young students can engage in the reasoning and decision-making needed for planning a solution for engineering design problems. Drawings, for the purposes of this study, are assumed to be representations of 1st-grade students' initial ideas for a solution to a presented engineering design problem. To that end, the portion of the study described in this chapter focused on how 1st-grade students' preliminary drawings of their proposed solutions related to the requirements of the presented engineering design problem and how their preliminary drawings related to their constructed artifacts.

The first question explores if 1st-grade students have resources for planning an idea—the ability to interpret the requirements of the problem, formulate a potential solution, choose between different materials, and represent their solution and choices via drawing—all prior to working with the materials. Students' initial reasoning and ideas are only productive in a planning context if they persist beyond the initial drawing to inform the artifact construction. Hence, the second question then looks to see how their drawing and artifact are related to understand if 1st-grade students carry an idea from drawing to artifact construction. Understanding the preliminary reasoning and decision-making that 1st-grade students are capable of and how their initial ideas are carried through the design process informs how we think about practices to engage young children in engineering design problems, particularly around planning.

PRIOR RESEARCH ON THE ROLE OF DRAWINGS AND ARTIFACTS IN THE ENGINEERING DESIGN PROCESS

The products of engineering can be seen everywhere—bridges, buildings, bubble gum, cell phones, cars, petroleum products, and more. They are the result of the defining activity of engineering: design. Engineering design is the process through which engineers address a problem or need through the creation of a product or process. Engineering design problems are typically ill-defined—meaning they have missing information, vague requirements, and multiple criteria for success (Jonassen, Strobel, & Lee, 2006)—because those facing or presenting the problems do not fully understand or even know all the variables and nuances that help explain a priori what would be the ideal solution. A diversity of models of design processes has been developed over time as a framework for people to solve ill-defined problems (Jonassen, 2000). These models (e.g., French, 1998; Pahl & Beitz, 1984) break the process into steps of problem definition/clarification, research, brainstorming, planning, prototyping, testing, and creating representations of a final solution. The purpose of the model is to help guide engineers in their thinking and actions as they clarify the problem, formulate possible solutions, create and evaluate solutions, and ultimately create drawings and models of their final solution.

While the creation of representations plays a major role at the end of the engineering design process to fully specify the final design, this study focuses on the act of creating representations in the early part of the engineering design process. Prior to prototyping a design idea, engineers engage in brainstorming and planning their initial idea, often representing it via drawing. Prior studies have theorized that adults engaged in design use preliminary drawings as external storage devices (Purcell & Gero, 1998; Simon, 1996) for elements of their design so they can continue to think about other requirements or constraints as they work to specify their initial idea for prototyping. This externalization of ideas via drawing is thought by many researchers to be essential to the process of design. This view is substantiated by a selection of research (Schütze, Sachse, & Römer, 2003; Song & Agogino, 2004) that has found that the amount of time spent engaged in drawing and/or the quality of preliminary drawings is related to the quality of the final solution. However, case studies and other empirical work (Bilda, Gero, & Purcell, 2006; Yang & Cham, 2007) have found no relationship between the amount and quality of initial drawings and the outcome for design problems—suggesting that mental imagery and visualization skills could be sufficient for many professionals without the need for preliminary drawings.

As engineering is introduced at the K–12 level, simplified models of the engineering design process (e.g., "Engineering is Elementary," 2012; Massachusetts Department of Education, 2001) have been created. In these K–12 models, the essential activities of the engineering design process that professional engineers use are preserved, particularly the element of planning via drawing. In practice, K–12 students are often tasked with planning by being required to create a drawing of their idea for their solution to an engineering design problem prior to constructing it. However, similar to adults, there is no consensus in the engineering education community as to whether planning, and specifically planning via drawing, is accessible and/or useful to K–12 students.

Research looking at how young children engage in planning solutions to a general problem-solving task shows that children as young as 7 years of age can plan solutions to problems, that they are sensitive to the contextual need for planning, and that planning impacts the success of their solution. Gardner and Rogoff (1990) studied children's ability to plan routes when presented with mazes and different requirements. They found that children in the range of 7- to 10-year-olds planned less often when they were asked to complete a maze as quickly as possible. However, when the children were presented with solving a maze as accurately as possible (making the fewest wrong turns), they were more likely to plan. Gauvain and Rogoff (1989) also found that 7- to 10-year-olds who planned routes for a model grocery store shopping trip created more efficient routes, suggesting that young children's planning influences the quality of their solution.

While research in general problem solving indicates that young children can plan, research looking at children engaged in open-ended design problems finds that children do not naturally plan. Johnsey's (1995) classroom observations of 4- to 10-year-olds engaged in design problems in a classroom context found that students "Make Evaluate Make" and did not spontaneously engage in planning (via drawing or otherwise). Similar results have been found among older students (10- to 13-year-olds) who are new to engaging in design problems around planning (McCormick, Murphy, & Hennessy, 1994; Welch & Lim, 2000), with the conclusion that planning is unnatural for novice designers who lack experience with manipulating the materials. The studies suggest that it may not be useful to require students to plan. However, with evidence that even very young children are sensitive to the contextual need for planning (Gardner & Rogoff, 1990), the conclusion that planning is unnatural for novice designers seems too general. It seems more likely that certain conditions may not create a contextual need for planning for students. For example, conditions that provide unlimited access to materials or where materials are entirely unfamiliar may make planning seem unnecessary or unhelpful to students.

A few studies have looked specifically at outcomes where students were required to create planning drawings prior to construction. Rogers and Wallace (2000) found no relationship between 5-year-olds' planning drawings and their final constructed artifacts. However, the researchers' comparison of drawings and artifacts was based on their qualitative observations and not well described. Their comparisons of drawing to artifact did not include any attempts to quantify the relationship between the two. Hence, from their work it is impossible to determine if all the drawings and artifact pairs were completely different or if, for some students, there were similarities between the drawing and artifact. In contrast, Fleer (2000) looked at 3- to 6-year-old children and categorized comparisons between artifact and drawing (e.g., naming of artifact; materials they intended to use; configuration and joining of materials; overall placement). She found that a majority of the artifacts that students constructed were related to their drawings, but that their construction ideas were easily altered by the proximity and availability of materials as well as other students' ideas. Her work suggests that planning may be within the grasp of young students.

One of the aims of the study I will describe in this chapter was to mitigate some of the challenges that surround looking at students' planning in classroom settings, by instead looking at students in individual interview sessions. Moreover, another aim was to establish clearer connections between the required preliminary drawing and the problem posed to students as well as connections between the preliminary drawing and final constructed artifact.

HOW THE STUDY WAS DESIGNED AND CARRIED OUT

Thirty-one 1st-grade children (16 males and 15 females) from an upper-middle-class suburban elementary school (7% free or reduced-price lunch, 72% Caucasian) outside of Boston, Massachusetts, participated in individual videotaped interviews with me. The interviews were conducted in the first half of the school year during the school day, and participants had not had any formal engineering experiences in school at that time. The participants were presented with a single engineering design problem, "The Trapped Keys." The problem required them to design and construct a tool to retrieve a set of two keys attached to a key ring from the bottom of a clear Plexiglass box (23 inches x 4.5 inches x 4.5 inches; see

Figure 11.1). Students were provided with nine materials (12-inch pipe cleaners, 12-inch wooden dowel rods [sticks], plastic spoons, unsharpened pencils, string, paper clips, round and bar magnets, clothespins, and tape) as well as scissors to use to design and construct a tool to retrieve the keys.

The materials presented to students were all selected as craft or everyday materials with which students were expected to be familiar. The goal of presenting students with familiar materials was to focus on the task of designing and constructing a tool and not exploring the materials. The specific combination of materials was chosen to give students a combination of ideal and non-ideal materials (see Tables 11.1 and 11.2) for the task to allow for multiple solution paths as well as to explore how well students understood the problem requirements of length (to reach the bottom of the box) and key acquisition (to retrieve the keys from the bottom of the box). For example, students were presented with 12 inch dowel rods and 7 3/8-inch pencils. Understanding the requirement of length would make the selection of the longer dowel rods a more ideal choice. Similarly, while magnets and spoons could both be used to pick up keys off the floor, understanding how the narrow box would make it difficult to get the spoons under the keys (whereas magnets can pick up the keys from multiple angles or positions) would make the magnets a more ideal choice.

During the interview session, each participant was shown all the available materials by the interviewer and presented with the task ("Make a tool to get the keys out of the box") in the following way:

> Let me tell you what we are going to do with all this stuff. It's got two parts to it. It's got a drawing part and a making part. So, you can see I've got some keys trapped in the bottom of this box. Kind of a little dangerous box. It's got some sharp edges and stuff. So, I want you to see if you can build me something that will get the keys out of the box without putting your hand in the box or moving the box. And before you build it I'm going to ask you to draw a picture of what you think you are going to make. Let's look at all the stuff you've got here so you know what you have to work with. You have pipe cleaners, super-long sticks, string, obviously scissors so you can cut stuff or do whatever you want with them, all kinds of magnets, tape. . . . And you can use any of these materials any way you want. And if you need help cutting or tying or taping you can ask me for help.

When the drawing was complete, the interviewer labeled the materials in the picture with confirmation from the participant. Participants were not informed about the properties of any of the materials and were not permitted to experiment with the materials prior to drawing. Once the drawing was completed, each participant worked on the construction of his or her solution. Children were not required to create a new drawing if they changed their first idea. After 10 minutes

Figure 11.1. Trapped Key Task and Materials

Table 11.1. Ideal and Non-Ideal Materials for Length

Ideal Materials for Length	Non-Ideal Materials for Length
Sticks—12 inches long	Pencils—Usable but shorter than sticks
Pipe Cleaner—12 inches long	Spoons—Difficult to attach together
String—As long as needed	

Table 11.2. Ideal and Non-Ideal Materials for Key Acquisition

Ideal Materials for Key Acquisition	Non-Ideal Materials for Key Acquisition
Magnets—Attracted to keys	Clothespin—Difficult to manipulate in box
Paper Clip—Can be bent to hook of small diameter	Tape—Difficult to stick to keys (rough surface). Attracted to side of box (static)
Pipe Cleaner—Can be bent to hook of small diameter	Spoons—Difficult or impossible to get under keys in narrow box

of construction time, the interviewer intervened to work with students who had not been able to create a successful solution. Together, interviewer and student constructed a solution that retrieved the keys so that all students felt they had been successful. The amount of assistance provided to students varied. Some students needed help connecting materials to make their own idea functional while others needed a completely new idea for a solution. The degree of assistance was not evaluated for this study because of time constraints, imposed by conducting the interview during the school day, prevented the implementation of a consistent protocol for providing assistance.

HOW THE CHILDREN'S DRAWINGS WERE INTERPRETED

To examine what requirements of the problem (length and key acquisition) students had attended to, and to examine their choice of material, each drawing was evaluated for the criteria shown in Table 11.3.

Figure 11.2 shows an example of the scoring for a drawing of a tool that received the highest score in each category (i.e., length and key acquisition). The participant's drawing included materials to address the requirements of length and key acquisition (two ideas for key acquisition) and selected ideal materials (string for length, and pipe cleaners/magnets for key acquisition) for both requirements. String is evaluated as sufficient material for length as it can be cut to lengths that reach the bottom of the 23-inch box.

Figure 11.3 shows a drawing that includes only a string (an ideal material for length that could reach the bottom of the box) with no material included that could be used for acquiring the keys.

Table 11.3. Scoring of Drawings

Category	Scoring
Addresses LENGTH	(1 point) One or more materials shown that would achieve length (0 points) Not Addressed/Unclear
Enough Material(s) for LENGTH	(2 points) Sufficient materials (2 or more materials that could reach the 23 in. to the bottom of the box) (1 point) Partial materials (too short) (0 points) No materials/Unclear
Ideal Materials for LENGTH	(2 points) All ideal materials (1 point) Some ideal materials (0 points) No Ideal Materials/Unclear
Addresses KEY ACQUISITION	(1 point) One or more materials that could acquire keys (0 points) Not addressed/Unclear
Ideal Materials for KEY ACQUISITION	(2 points) All ideal materials (1 point) Some ideal materials (0 points) No ideal materials/Unclear

Figure 11.2. A Student's Drawing of His Planned Solution That Addresses Length and Key Acquisition

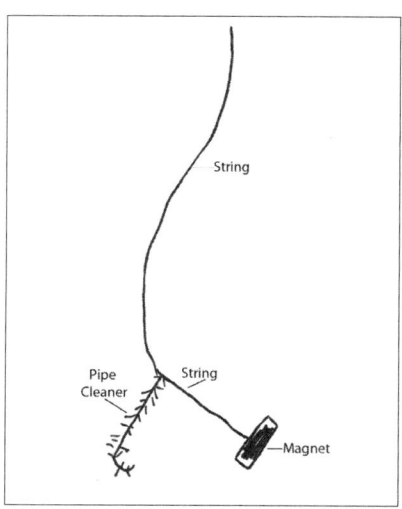

Category	Score
Addresses LENGTH	1 (Addressed)
Enough Material(s) for LENGTH	2 (Materials able to reach 23 inches or greater)
Ideal Materials for LENGTH	All Ideal Materials
Addresses Key Acquisition	1 (Not Addressed)
Ideal Materials for Key Acquisition	2 (All Ideal Materials)

ANALYSIS OF RELATIONSHIP BETWEEN DRAWING AND ARTIFACT

Examination of students' drawings and their final constructed artifacts could be characterized by a range of relationship types—from artifacts that were identical to the drawing to those bearing no resemblance to the drawing. To better understand the nature of the relationship, a Total Relationship score was calculated by comparing the materials, their arrangement, and quantity used for both length and key acquisition in the drawing and artifact. Total Relationship scores were only calculated for the final artifacts that successfully retrieved the keys in the designated time. Students moved quickly through intermediary artifact construction, making it difficult to identify what was an intermediary artifact that could be compared to the drawing. For students who were not successful, they were in various states (no artifact, partial artifact, nonfunctional artifact) at the conclusion

Figure 11.3. A Student's Drawing of Her Planned Solution That Addresses Only Length

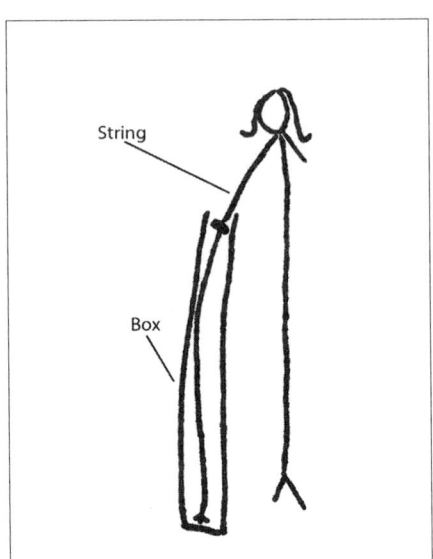

Category	Score
Addresses LENGTH	1 (Addressed)
Enough Material(s) for LENGTH	2 (Materials able to reach 23 inches or greater)
Ideal Materials for LENGTH	2 (All Ideal Materials)
Addresses Key Acquisition	0 (Not Addressed)
Ideal Materials for Key Acquisition	0 (No Ideal Materials/Unclear)

of independent building (10-minute mark). Hence, the final artifact of successful students was the most consistent artifact state to compare with the initial drawing. Table 11.4 shows the categories and calculations for the Total Relationship score for the final artifact and drawing.

A Total Relationship score of 6 would represent a drawing/artifact pair where the artifact was identical to the drawing. A score of 0 would represent a drawing/artifact pair where the final artifact looked nothing like the original drawing. In Figure 11.4, which has a Total Relationship score of 6, the artifact is an exact match to the original drawing in terms of the materials selected and their attachment for length and key acquisition.

Figure 11.5 has a Total Relationship score of 2. Tape, not shown in the drawing, was added to the artifact to attach the stick and pipe cleaner, which yielded a

Table 11.4. Total Relationship Score—Relationship Between Drawing and Artifact Scores

Length Relationship Score in Drawing and Artifact (LRS)	3 points—Identical
	2 points—Differ in attachment material or different attachment position or increase/decrease of material quantity shown in drawing
	1 point—Differ in material arrangement or additional material types used
	0 points—Completely different
Key Acquisition Relationship Score in Drawing and Artifact (KARS)	3 points—Identical
	2 points—Differ in attachment material or different attachment position or increase/decrease of material quantity shown in drawing
	1 point—Differ in material arrangement or additional materials types used
	0 points—Completely different
Total Relationship Score (TRS)	LRC + KARS

Length Relationship score of 2 for the pair. The material for key acquisition was completely changed from tape to a magnet (attached with paper clips), yielding a Key Acquisition Relationship score of 0.

It should be noted that the Total Relationship score was only calculated for successful solutions (solutions that retrieved the keys in under 10 minutes) based on the final artifact that was used to retrieve the keys from the box, since incomplete solutions did not have an equivalent artifact that could be evaluated.

RELATIONSHIPS BETWEEN CHILDREN'S DRAWINGS AND ARTIFACTS

How 1st-grade students' preliminary drawings of their proposed solutions related to the requirements of the presented engineering design problem was addressed by analyzing students' drawings as they related to the problem requirements and students' material selection. The results of the analysis, summarized for all student drawings in Table 11.5, show that the majority of 1st-grade students were able to engage in significant thought and action about the problem. The drawings, which were completed prior to students' manipulating the materials, illustrate that students made decisions about the nature of the tools and optimal materials based on the problem requirements and materials available.

Figure 11.4. A Drawing and Artifact Pair with a Total Relationship
Score of 6 (LRS = 3; KARS = 3)

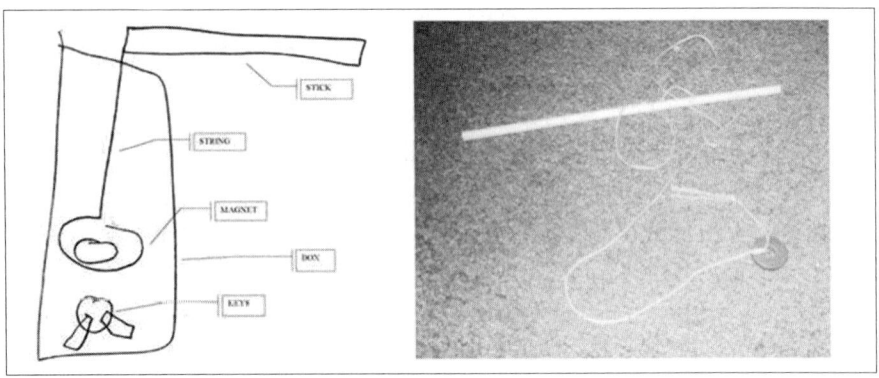

Figure 11.5. A Drawing and Artifact Pair with a Total Relationship
Score of 2 (LRS = 2; KARS = 0)

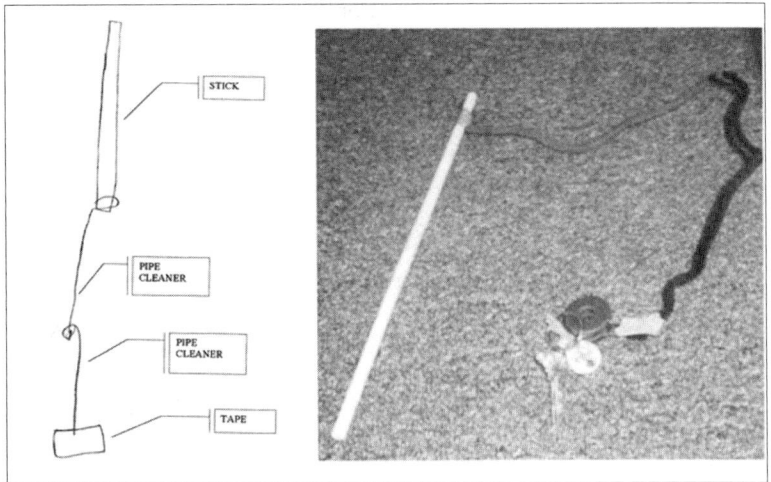

Table 11.5. Frequencies and Percentages of Students' Drawings That
Addressed or Not Each of the Problem Criteria (All Participants, N = 31)

	Addressed	One or More Ideal Materials	Sufficient Materials
Length	28 (90.3%)	26 (83.9%)	16 (58.1%)
Key Acquisition	25 (80.6%)	17 (51.6%)	N/A

The majority of 1st-grade students (80.6%) could extract from the statement of the problem ("Make a tool to get the keys out of the box") that they needed to create something long and that it would need to have something to retrieve the keys. Length was addressed in participants' drawings with at least one material by 90.3% of the students, with 83.9% of them addressing it with one or more ideal material. In addition, students were able to evaluate the materials available and select ideal ones for both length and key acquisition. Most of the students' drawings (83.9%) included one or more ideal materials for length. Likewise, 54.8% of the drawings included one or more ideal materials for key acquisition. There is a difference in the percentage of students' drawings that included ideal materials for length (83.9%) and key acquisition (54.8%), suggesting that students may have had more ideas about the possible best materials for length than for key acquisition. The quantity of materials included in the drawing was evaluated for length and 58.1% of the students included multiple materials (e.g., pipe cleaner and stick) or a single material (string) that could achieve a length of 23 inches (or longer). This is seen as evidence that many students were able to think specifically about the dimensions of the box and how materials could be combined to achieve the necessary length.

I considered how 1st-grade students' preliminary drawings related to their constructed artifacts by analyzing how closely students' final artifact resembled their initial drawing. Table 11.6 shows the distribution of the results of the analysis of the relationship between students' drawings and artifacts.

More than half the students (58.3%) had a high Total Relationship score, indicating that their artifact was identical or very similar to their drawing. Only one student had an artifact that had no discernible relationship to his drawing. The persistence of students' ideas from drawing to artifact suggests that these initial ideas are significant and can inform artifact construction.

PLANNING IN CHILDREN'S ENGINEERING EDUCATION EXPERIENCES

Previous research left open the question of how and if children, particularly those as young as 1st-grade students, can engage in planning in the context of an engineering design problem. The preliminary drawings created by the students in this study show that many of the young student participants were able to engage in significant reasoning and decision-making about the problem presented to them prior to engaging in any hands-on manipulation of the materials. The majority of 1st-grade students in this study were able to take the problem ("Make a tool to get the keys out of the box") and infer the requirements of length (the tool had to be long enough to reach the bottom of the box) and key acquisition (a means of

Table 11.6. Frequencies and Percentages of Students Who Received Each Category of Total Relationship Score (Only Successful Solutions Were Scored, *N* = 24)

Total Score Categories	Number of Students
High Total Relationship Scores (scores of 4–6)	14 (58.3%)
Low Total Relationship Scores (scores of 1–3)	9 (37.5 %)
No Relationship (scores of 0)	1 (4.2%)

obtaining the keys had to be designed). They also evaluated materials based on the requirements and selected materials accordingly. Moreover, in general, their initial idea to solve the problem, represented through their drawing, persisted (to different degrees) even when they began to work with the actual physical materials.

An acknowledged weakness of this study is the lack of explicit information about students' particular experience with the materials that were available to them. While students' experience with the materials was not evaluated, the materials were selected with the assumption that they would be familiar to students. This familiarity would allow students to be able to make informed decisions about their use to solve a novel problem, as engineers do when they select materials for a design. The drawings support this assumption around the idea of length, as most students (83.9%; see Table 11.5) choose ideal materials (12-inch dowels or pipe cleaners) over non-ideal materials (shorter pencils and spoons). This suggests that they were familiar enough with the materials to make decisions when evaluating materials for the requirement of length. However, students choose a greater variety of materials (ideal and non-ideal) for key acquisition. This could be because they had less experience with the materials in the context of the requirement of key acquisition (retrieving something from a small space). In all likelihood, some students were probably learning about the key acquisition materials once they began construction, learning about their properties for the purpose of acquiring the keys. If students had greater experience with the key acquisition materials, it would seem only to strengthen the results by allowing students to make more informed choices when drawing.

Overall, these results are supportive of the idea that planning (preliminary idea generation and material selection) may be within the grasp of many 1st-grade students. It should be emphasized that the results of this study do not allow us to say that the students engaged in this study were planning via drawing. This study did not provide evidence that the students used the drawing as an external storage or reflection tool as adult engineers do (Purcell & Gero, 1998; Simon, 1996) or that students even use the drawings to inform their construction. However, the drawings in this study are evidence that students can engage in the kind of preliminary reasoning and decision-making that would be useful in planning—1st-grade students can make design choices prior to manipulating the materials.

While planning may be within the grasp of 1st-grade students, it is unclear whether drawing may be the tool that facilitates planning. The fact that over half of the students (58.3%; see Table 11.6) had drawings that were nearly identical to their final constructions raises the question of whether 1st-grade students may create internal (mental) representations of their ideas that they are simply reproducing in the form of a drawing and a constructed artifact. Further exploration of 1st-grade students' potential methods of planning is needed to better understand what tools, resources, or strategies would be appropriate to include in engineering design activities for young children.

REFERENCES

Bilda, Z., Gero, J. S., & Purcell, T. (2006). To sketch or not to sketch? That is the question. *Design Studies, 27*, 587–613.

Engineering Is Elementary. (2012). Retrieved from http://www.mos.org/eie/

Fleer, M. (2000). Working technologically: Investigations into how young children design and make during technology education. *International Journal of Technology and Design Education, 10*, 43–59.

French, M. J. (1998). *Conceptual design for engineers*. London, UK: Springer.

Gardner, W., & Rogoff, B. (1990). Children's deliberateness of planning according to task circumstances. *Developmental Psychology, 26*, 480–487.

Gauvain, M., & Rogoff, B. (1989). Collaborative problem solving and children's planning skills. *Developmental Psychology, 25*(1), 139–151.

International Technology Education Association. (2002). *Standards for technological literacy: Content for the study of technology*. Reston, VA: International Technology Education Association.

Johnsey, R. (1995). *The place of the process skill making in design and technology: Lessons from research into the way primary children design and make*. Paper presented at the IDATER95: International Conference on Design and Technology Educational Research and Curriculum Development, Loughborough, UK: Loughborough University of Technology.

Jonassen, D. H. (2000). Toward a design theory of problem solving. *Educational Technology Research and Development, 48*(4), 63–85.

Jonassen, D. H., Strobel, J., & Lee, C. B. (2006). Everyday problem solving in engineering: Lessons for engineering educators. *Journal of Engineering Education, 95*(2), 139–151.

Massachusetts Department of Education. (2001). *Science and technology/engineering curriculum framework*. Retrieved from http://www.doe.mass.edu/frameworks/scitech/1006.pdf

McCormick, R., Murphy, P., & Hennessy, S. (1994). Problem-solving processes in technology education: A pilot study. *International Journal of Technology and Design Education, 4*, 5–34.

Pahl, G., & Beitz, W. (1984). *Engineering design*. New York: Springer-Verlag.

Purcell, A. T., & Gero, J. S. (1998). Drawings and the design process. *Design Studies, 19*, 389–430.

Rogers, G., & Wallace, J. (2000). The wheels of the bus: Children designing in an early years classroom. *Research in Science & Technology Education, 18*(1), 127–135.

Schütze, M., Sachse, P., & Römer, A. (2003). Support value of sketching in the design process. *Journal of Engineering Design*, 89–97.

Simon, H. (1996). *The sciences of the artificial* (3rd ed.). Cambridge, MA: MIT Press.

Song, S., & Agogino, A. M. (2004). *Insights on designers' sketching activities in product design teams.* Paper presented at the ASME Design Engineering Techncial Conference, Salt Lake City, Utah.

Welch, M., & Lim, H. S. (2000). The strategic thinking of novice designers: Discontinuity between theory and practice [Electronic Version]. *The Journal of Technology Studies, XXVI*. Retrieved January 15, 2008, from http://scholar.lib.vt.edu/ejournals/JOTS/Summer-Fall-2000/welch.html

Yang, M. C., & Cham, J. G. (2007). An analysis of sketching skill and its role in early stage engineering design. *Journal of Mechanical Design, 129*, 476–482.

CHAPTER 12

From Seeing Points to Seeing Intervals in Number Lines and Graphs

Analúcia D. Schliemann,

David W. Carraher, and Mary C. Caddle

From bar graphs to graphs of linear and nonlinear functions, teaching and learn-ing about graphs pervades the K–12 mathematics curriculum. Even though func-tion graphs are only introduced in the middle school years, research shows that, with suitable instruction, elementary school children can learn about variables and functions and can use multiple systems of representation for functions, including graphs (Carraher & Schliemann, 2007). This chapter describes partial results of an early algebra longitudinal intervention with 3rd- to 5th-grade students where, building on initial work with number lines in 3rd grade, 4th-graders were intro-duced to the Cartesian space by first representing relations on parallel number lines and then setting these at right angles to build the Cartesian plane. Here, we describe the intervention and report on its impact on 5th-graders' interpretation of events or quantities represented along intervals on the real number line and in graphs of linear functions relating time and distance.

NUMBER LINES AND FUNCTION GRAPHS IN MATHEMATICS AND IN EDUCATION

To successfully interpret graphs of functions students need to understand how the geometrical properties of graphs (highest and lowest points, slope, curvature, intersections, and so on) represent the structure and behavior of functions (kind of function, local minima and maxima, rates of change, co-variation, solution sets, inequalities) as well as the situations the functions are intended to model. Graphs of functions in the plane represent relations of elements along the *x*- and *y*-axes, each of which can be regarded as a real number line.

Understanding graphs of functions therefore requires a solid understanding of the real number line as a representation of numbers and of intervals between numbers. However, in middle school and even high school, when students learn about graphs of functions, the individual points corresponding to the intersection of two projection lines, one perpendicular to the x-axis, the other perpendicular to the y-axis, become the prominent feature and more than often the interval representation is ignored (see Bell & Janvier, 1981). This is not surprising given that, in elementary school, instruction on graphs is usually restricted to bar graphs and point readings.

The number line can be seen as a physical object or as a display of simple mathematical objects, possibly to convey information on everyday events or scientific situations represented by the number line. It constitutes a representational system with its own properties, as well as conventions about usage and interpretation that allow communication about events and mathematical ideas. Bass (1998) has argued for an early emphasis on the geometric real line model of real numbers in mathematics education. Accordingly, number lines have been introduced in the early grades, in traditional form (Diezmann & Lowrie, 2006; Fuson, 1984) or in the form of an empty number line (Gravemeijer, 1993; Gravemeijer & Stephan, 2002; Klein, Beishuizen, & Treffers, 1998; Treffers, 1991), mainly as a tool to help children learn about addition, subtraction, and the decimal number system. Treffers (1991) exemplifies how parallel empty number lines can also help students meaningfully learn about fractions, ratios, and proportions. Thus, the number line can play major roles throughout K–12 mathematics education, from early work with numbers to the introduction of algebraic notation and beyond. It provides a consistent backdrop for a wide range of critical topics in arithmetic, geometry, and algebra. Of special importance for this chapter is the fact that two real number lines set at right angles can eventually be used to represent functions in the Cartesian plane.

Our view has been that, even in elementary school, students need to shift the focus from individual points to intervals along the number lines constituting the axes of the plane and to segments of the graph. This would allow them to detect information not held by individual points, a crucial step in understanding function graphs (Bell & Janvier, 1981; Kosslyn, 1989; Leinhardt, Zaslavsky, & Stein, 1990). However, research on students' work with number lines shows that we cannot assume that they appropriately draw upon intervals in conceptualizing problems represented along number lines (see Carraher, Schliemann, Brizuela, & Earnest, 2006). When questioned about the differences between two values on a number line, young children may look for entities to count. To determine the difference between two numbers, some students may count notches (fenceposts) at integer values on the number line (Gravemeijer & Stephan, 2002; Treffers, 1991). Counting notches may or may not produce a correct answer, and the student may be unsure about whether notches at the ends of the interval should be included. At issue is a major shift in the meaning of number from counts to measures (Freudenthal, 1983): Mastery of number lines, as well as of graphs in the Cartesian space, requires suitably employing numbers both as position points and as magnitudes of intervals. Children need to learn about the number line in ways that stress the interval feature of the number lines.

The Intervention and the Evaluation of Its Impact

Some of the activities in our early algebra intervention focused specifically on how 3rd- through 5th-grade students were introduced to the number line and to graphs in the Cartesian space. One of our goals was to prepare them to understand graphs by considering and interrelating intervals in the graphs of linear functions, instead of focusing on isolated points.

The impact of the intervention was evaluated through a written assessment given to intervention and control group students in 5th grade, at the end of the intervention. We also analyzed the strategies and challenges faced by the intervention group students only as they, in an interview, answered questions designed to clarify their written responses about a graph of a linear function. Our analysis aimed at identifying (a) the impact of the intervention on students' written assessment performance, as compared with a control group, on questions about the representation of numbers and events through number lines and graphs, and (b) intervention students' strategies and challenges as they, in the interview, attempted to interpret a linear function graph and determine how intervals on one axis related to intervals on the other.

The 3-year longitudinal classroom intervention with 3rd- through 5th-grade students took place in an inner-city public school in Boston, Massachusetts, located in a predominantly African American community with a strong presence of immigrants from Cape Verde and the Azores Islands.

In the first year of the intervention, all 22 students in the school's two 3rd-grade classrooms (ages 8 to 9) participated in two 60-minute weekly algebra lessons, led by members of the research team, and two 30-minute reviews of homework, led by their regular schoolteacher. The same students participated in our intervention the following year in 4th grade, using the same format. In 5th grade, they participated in one 90-minute lesson each week, followed by 45 minutes of homework review. At the end of each year, each student was given a 50-problem written assessment; seven of these problems, on number lines and graphs, are analyzed here. The 21 students also participated in individual interviews, led by members of the research team, on some of their written assessment responses. Written assessment data were also collected from a control group of 24 students in the same school, who attended 3rd through 5th grades the immediate prior years.

The Early Algebra Lessons

The lessons focused on algebra as a generalized arithmetic of numbers and quantities and aimed at promoting a shift from computations on particular numbers and measures toward thinking about relations among sets of numbers. Problem contexts constituted ways to situate and deepen learning and generalizations about quantities and numbers. Lessons included topics in the elementary school curriculum such as addition, subtraction, multiplication, division, fractions, ratio, and proportion, addressed in an "algebrafied" manner (Kaput, 1998) and linked to variables and linear and nonlinear functions. Central to our approach was the use of multiple representations, namely, natural language, function tables, number lines, Cartesian graphs, algebraic notation, and equations, with number lines being introduced in 3rd grade. Our lessons are available at www.earlyalgebra.org and some of them are described in detail by Carraher and Schliemann (2007), Carraher et al. (2006), Carraher, Schliemann, and Schwartz (2008); Peled and Carraher (2008), Schliemann and Carraher (2002), and Schliemann, Carraher, and Brizuela (2007). The lessons on number lines and graphs are outlined below.

Working with Number Lines

Instruction on number lines focused on intervals, as opposed to points, and led to the Cartesian space by first working with the representation of relations in parallel number lines and later rotating one of them to become perpendicular to the other, thus allowing for representation of relations as graphs.

We started by using a piece of twine with numerals for successive integers attached at 1-foot intervals to initiate discussions with 3rd-graders about the number line. Students at first tended to consider only the numerals shown on the twine. However, after discussing where the number line would finish, they quickly realized that the number line is not a physical object confined within the classroom walls but is rather an imaginary construct that can "go through walls" and continue indefinitely in two directions.

In the three following lessons, we introduced number lines drawn on paper and projected onto a screen. To emphasize intervals, as opposed to points on the number line, arrows linking points on the same line were proposed to represent changes in values, and longer arrows connecting shorter ones were used to lead students to consider shortcuts that went from the tail of the first arrow to the head of the last arrow (see Figure 12.1).

Students worked on expressing shortcuts or simplifications through paper strips and through notation. For example, the operation "+ 7 −9" could be represented as subtracting 2 since each expression had the same effect. Students also used number lines to model additive relations in word problems by displacing themselves along a number line drawn on the classroom floor or by showing displacements on paper.

Figure 12.1. Number Line Intervals and Shortcut

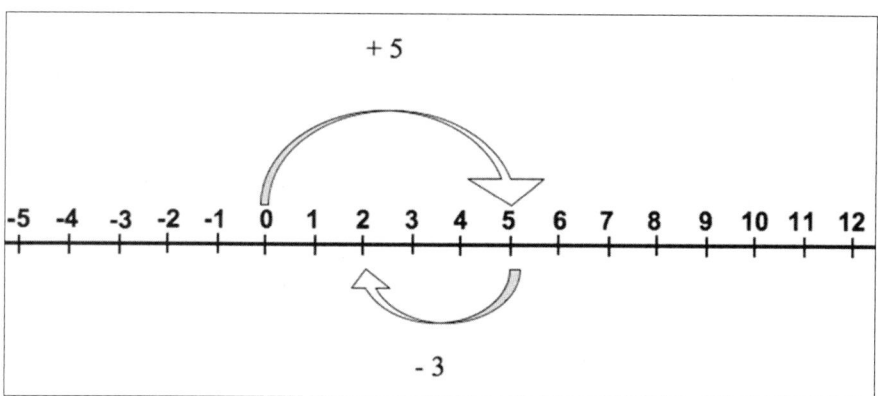

In the fifth 3rd-grade lesson on number lines, we introduced the "*N*-number line"—a variation on a standard number line in which (see Figure 12.2) positions were identified as $N - 2, N - 1, N, N + 1, N + 2$, and so forth. Students were asked to express additive operations on the *N*-number line to solve addition and subtraction problems where the starting point had no determined value, thus focusing on variables and notation for variables.

From Parallel Number Lines to the Cartesian Space

The transition from single number lines to axes in the Cartesian space was carried out near the end of 3rd grade, first using the classroom floor as a plotting surface (see Schliemann & Carraher, 2002, for details on a similar lesson with another group of students). Two parallel number lines were constructed on the floor using tape and magic marker.

In the Cartesian space, each point on an axis is associated with all points on a perpendicular projection to the axis at that value. To address this feature, we first worked with a single number line, emphasizing that one does not need to be on top of the line itself to represent a given value. Figure 12.3 shows a child exploring places representing 9 hours that were located "off" the number line drawn on the floor, as he moves along an imaginary perpendicular line that crosses the drawn line at 9.

Students were asked to represent the relationship "for each hour of work one gets \$2" on two parallel number lines, one standing for hours worked, the other for number of dollars earned. The role of each student was to maintain contact with the appropriate positions on each number line, a task that became increasingly difficult and ultimately impossible as the amount of hours worked increased. Figure 12.4 shows a student attempting to represent, at the same time, 5 hours and \$10 on the two parallel number lines.

Figure 12.2. The *N*-Number Line

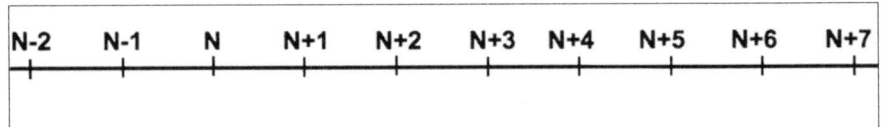

Figure 12.3. A Student Backing Away from the Number Line Along Imaginary Projection Line for "9 (Hours)"

Figure 12.4. A Student Attempting to Show That She Worked 5 Hours (Right Foot on a Point on a Virtual Line Perpendicular to Lines on the Floor, Crossing the Point Marked 5 on the Line Farther from Her) and Earned 10 Dollars (Left Foot Corresponding to 10 Dollars on the Line Closer to Her).

By rotating the earnings number line until it was perpendicular to the hours-worked axis, we introduced the Cartesian plane to simultaneously represent two values through a single point, with number lines corresponding to the *x*- and *y*-axes. For each pair of values (hours worked and earnings), students were asked to move to the appropriate point in the plane. At first, they tended to show one value on one axis and then walk to the other axis to show the corresponding value. However, with guidance, after no more than 20 minutes, they learned how to make adjustments in one variable without altering the value of the other, and stand on the point representing a pair of values (see Figures 12.5 and 12.6).

In subsequent lessons, students worked with word problems associated with functions of direct proportionality, making graphs on the blackboard and on paper. The transition from the classroom floor to paper and pencil appeared to go smoothly for the children. They interpreted, built, and discussed function tables and graphs to represent relations between payment per hour, candy bars per person, and distance per time.

The Assessment and Interview Problems

A 50-item written assessment was given at the end of each year to students in the intervention group and to a control group made up of 24 students in the same school who had attended 3rd through 5th grades the immediate prior years. Students were given two 60-minute periods to answer all the questions. The assessment included problems designed by the research team of mathematics educators and questions for grades 4, 6, and 8 from the National Assessment of Educational Progress (NAEP) and from the Massachusetts Comprehensive Assessment System (MCAS) tests. Design and choice of problems by the research group was based on their relevance to the assessment of learning of functional relations, algebraic notation, number lines, graphs, and equations.

Eight of the problems directly related to the representation of intervals or points on the number line and on graphs. Results on seven of these eight problems were used here for comparing intervention students' performance to that of control students. Of those, three problems (Figure 12.7) focused on representations of points or intervals on number lines, two (Figure 12.8) on the representation of points in the Cartesian plane, and two on the representation of linear functions in the Cartesian plane (Figures 12.9 and 12.10). The remaining problem required extending the lines of two linear functions to determine their point of intersection. Imprecise use of rulers by both groups prevented clear data analysis for this problem, so it was not included in this analysis.

A last problem, on a distance-time graph (see Figure 12.11), was given to the intervention group only. This problem was also the focus of an interview that took place a week after the written assessment. Answers to this problem, on each occasion, allowed for analysis of intervention students' strategies and of the challenges they face in interrelating intervals in the Cartesian space.

Answers to each question were categorized as correct or incorrect by two judges, always with more than 80% agreement. Disagreements were solved through discussion between the first two judges and consultation with a third judge. The same process was used for determining the amount of help given to students in the interview and for judging whether or not the students focused on intervals or on specific points, as they answered questions interrelating intervals.

Figure 12.5. Showing That You Worked 5 Hours and Received 10 Dollars

Figure 12.6. Three Students (Boys in Center of Photo) Representing Three Coordinate Points for the Relation "2 Dollars for Each Hour of Work"

Figure 12.7. Three Written Assessment Problems on Number Line Representations

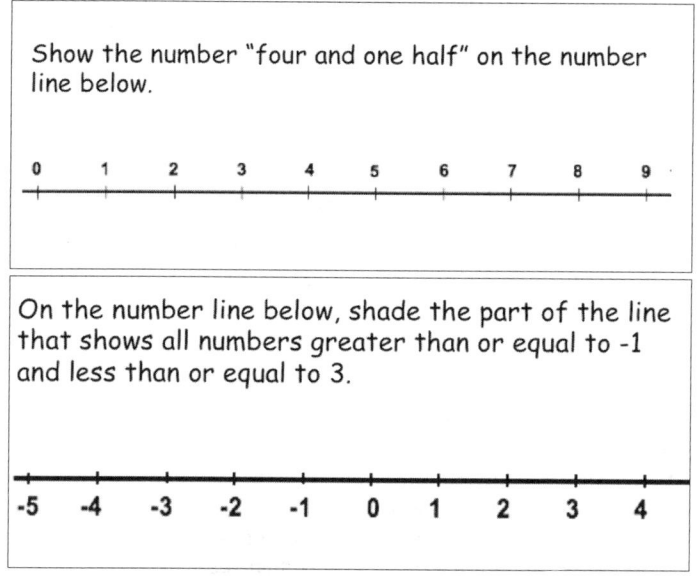

Show the number "four and one half" on the number line below.

On the number line below, shade the part of the line that shows all numbers greater than or equal to -1 and less than or equal to 3.

Bob started with K marbles.
He lost 3 marbles in a <u>first game</u>.
He then won 5 marbles in a <u>second game</u>.
Finally, he lost 4 marbles in the <u>third game</u>.

On the number line:
    Show where he started.
    Show where he was after each game.

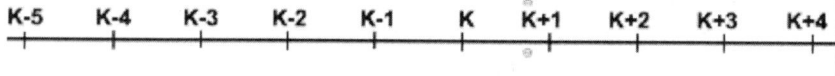

How many marbles did Bob have after he finished the third game?

Figure 12.8. Two Written Assessment Problems on the Representation of Points in the Cartesian Plane

The map below shows the location of some places of Keith's hometown

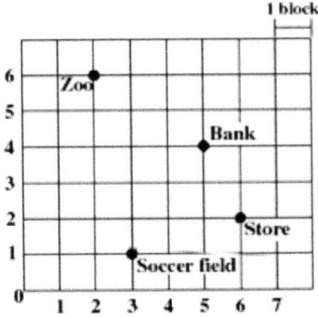

- What ordered pair names the location of the Bank?
- What is located at (2,6)?
- Moving along the grid lines, the shortest distance from the Store to the Bank is 3 blocks. Moving along the grid lines, what is the shortest distance from the Store to the Zoo?
- Moving along the grid lines, the shortest distance from the Library to the Soccer field is 7 blocks. What ordered pair could name the location of the Library?

Ms. Truman's students recorded the temperature at 2:15 p.m. each day for 6 school days. The data are shown below.

Monday   46°F

Tuesday   58°F

Wednesday 66°F

Thursday   49°F

Friday   38°F

Monday   50°F

Which graph below best represents the temperatures recorded from Monday to Monday?

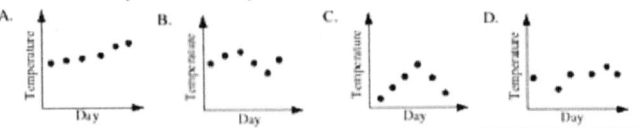

Figure 12.9. The First Written Assessment Problem on the Representation of Functions in the Cartesian Plane

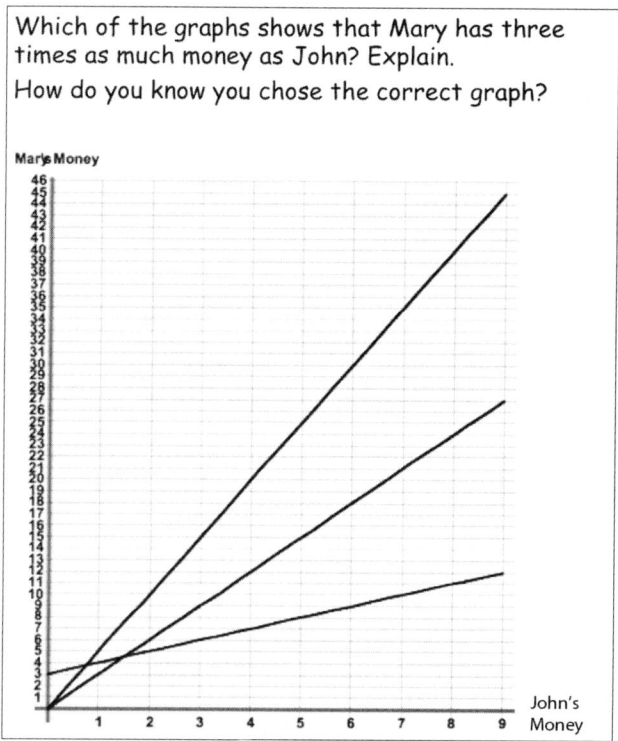

Which of the graphs shows that Mary has three times as much money as John? Explain.

How do you know you chose the correct graph?

Intervention and Control Group
Students' Written Assessment Performance

Table 12.1 shows the percentage of correct answers by the intervention and control groups in the written assessment problems.

The first number line problem asked students to mark the point 4.5 on the line, and the second asked to mark an interval, from -1 to +3. Differences between groups were not significant for these two questions. However, performance was notably different across questions. Whereas 86% of students in the intervention group and 100% of those in the control group located the point 4.5 on the line, only 32% of the intervention group and 25% of the controls correctly marked the interval from -1 to +3. This difference highlights the difficulties students have when considering intervals, rather than points, but it may also result from inclusion of a negative number as the starting point of the interval.

The third problem using number lines presented a story in which a player started with *K* marbles and then, over three successive games, lost three marbles, won five marbles, and lost four. Students were asked to show the problem on an *N*-number line labeled from $K - 5$ to $K + 4$, and then to determine how many marbles the player had at the end ($K - 2$). Here, the students in the intervention group did significantly better (Fisher exact probability=.011), with 64% of them correctly representing and solving the problem, while only 29% of those in the control group did so. This suggests that the intervention helped students use number lines as a tool for solving problems, even if they did not differ from the control group students in identifying points or intervals on the number line.

Of the four problems using the Cartesian plane, two (see Figure 12.8) did not result in significant differences between groups. These two problems asked students to look at specific points on the graph, rather than considering the graph as a whole. In the problem with points plotted in the plane as map locations and questions about the distances between locations and the coordinates of locations, the two groups performed equally well (42% and 43% correct answers). For the problem with a list of temperature data and a question about which discrete points

Figure 12.10. The Second Written Assessment Problem on the Representation of Functions in the Cartesian Plane

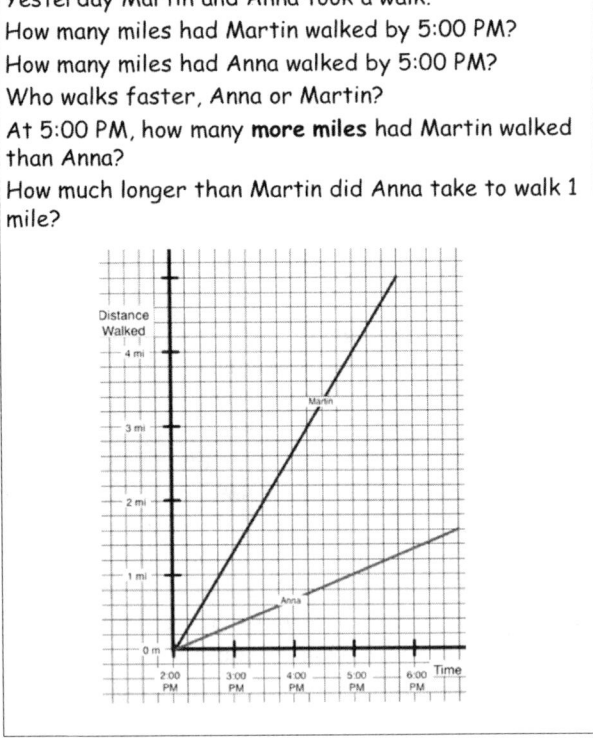

Figure 12.11. The Written Assessment and Interview Problem

WRITTEN ASSESSMENT PROBLEM:

The graph below shows the distance Jose walked depending on time. Answer the following questions:

(1) How many miles did Jose walk between 3 and 5 hours?

(2) How long did it take him to walk 10 miles total?

(3) How long did he take to walk from 10 to 25 miles?

(4) How fast (in miles per hour) did Jose walk?

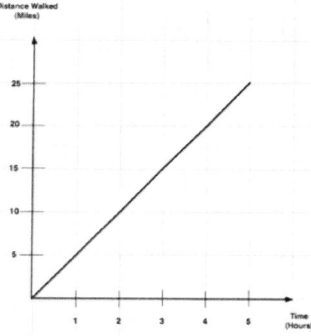

INTERVIEW QUESTIONS:

(1) Show on the graph the time between 3 and 5 hours and ask: In that time—from 3 to 5 hours— how many miles did Jose walk?

(5) Beginning at 4 hours, how much more time Jose will need to reach a total of 35 miles if he continues at the same pace? How do you know?

Table 12.1. Percentage of Correct Answers for Each of the Seven Written Assessment Items on Number Lines and Graphs

Problems	Intervention	Control	p
Representing 4.5 on the number line	86%	100%	NS*
Representing from –1 to 3 on the number line	32%	25%	NS
Representing and solving a marbles game on the number line	64%	29%	<.05
Library location point	42%	43%	NS
Matching temperature changes to points in a graph	82%	71%	NS
Choosing the graph that shows three times as much money	36%	13%	<.01
Comparing events in two linear function graphs	70%	56%	<.05

* NS: p value not significant

on the graph matched the data, the intervention group performed better (82% correct against 71%), but the difference between groups was not significant.

In contrast, the other two problems, which required considering intervals, led to significant differences between the experimental and control groups. The first problem (see Figure 12.9) showed a plane with three linear function graphs relating Mary's (*x*-axis) and John's (*y*-axis) amounts of money. Students were asked to choose which graph showed that Mary had three times as much money as John and to explain their choice. The intervention group performed significantly better ($p < .01$), with 36% correct answers against 13% for the control group. All control students explained their choice by mentioning only one point in the graphs (e.g., when John has one Mary has three), while six students (27%) in the intervention group considered the functional relation (e.g., for each dollar John has, Mary has three).

The second problem (see Figure 12.10) involved two graphs of distance as a linear function of time, one for Martin's walk, the other for Anna's. Students had to determine how many miles each person had walked, who walked faster, how many more miles one had walked than the other, and how long it would take someone to walk one mile. Again, the intervention group performed significantly better ($p < .05$), with 70% correct answers versus 56%.

The above results show that intervention students perform as well as their control peers in identifying points in the plane but show a significantly better understanding of items involving linear functional representations on the Cartesian space. Our next analysis explores how students who had participated in the intervention address the challenge of interpreting events represented in a linear function graph.

Students' Strategies and Challenges in Interpreting a Graph of a Linear Function

This second analysis focuses on the intervention group's strategies and challenges in interpreting the linear function graph representing a walk as a function of time and distance. This problem was included in the intervention group written assessment and in the follow-up interview (see Figure 12.11). Table 12.2 shows the percentages of correct answers to each question for the intervention students, in the written assessment and in the interview.

Question 1. In the written assessment, only four of the 22 students (18%) correctly determined the distance elapsed between 3 and 5 hours, namely 10 miles, with half of the students giving the same incorrect answer, 20 miles, to question 1. It appears that students interpreted "between 3 and 5 hours" as referring not to an interval but rather to a single value "between 3 and 5 hours"; thus, 4 hours, which would lead to their answer of 20 miles.

Table 12.2. Percentage of Correct Answers by the Intervention Group

Questions	Written Assessment	Interview
1 (*written assessment version*). How many miles did Jose walk between 3 and 5 hours?	18%	—
1 (*interview version*). From 3 to 5 hours, how many miles did Jose walk?	—	60%
2. How long did it take him to walk 10 miles total?	86%	—
3. How long did he take to walk from 10 to 25 miles?	45%	—
4. How fast (in miles per hour) did Jose walk?	59%	—
5 (*extra interview-only question*). Beginning at 4 hours, how much more time will Jose need to reach a total of 35 miles if he continues at the same pace?	—	80% (47% without help and 33% with little help)

To determine whether the lack of clarity in the written question affected students' responses, we reintroduced the question in the interview (see Figure 12.11, Interview Question 1), rephrasing it as "From 3 to 5 hours, how many miles did Jose walk?" With this revised wording, 60% of the students answered the question correctly. This suggests that their answers in the written assessment were due to the interpretation of "between 3 and 5 hours" as the single value between 3 and 5. However, four students persisted in selecting 4 hours on the x-axis and, through projection to the graph and then to the y-axis, arrived once more at the answer of 20 miles. For these students, wrong answers seem to result from a point-wise approach to the values along the graph axes, instead of considering the intervals in both axes. Another five students gave wrong answers, some due to computation mistakes, others for reasons that were not clear.

Question 2. In the written assessment, the majority of students (86%) correctly determined the time taken to walk a total of 10 miles. Here, the correct answer could be determined by simply reading off the time value (2 hours) associated with the elapsed distance of 10 miles from the origin, not explicitly considering intervals. Given the high number of correct written answers for this question, it was not included in the interview.

Question 3. In the written assessment, 45% of the students determined the elapsed time corresponding to the segment from 10 to 25 miles. Like Question 1, Question 3 asked students to consider a difference between two points and to relate a distance interval to a time interval. Better performance in Question 3 provides support to the assumption that wrong answers to assessment Question 1 resulted from the question's wording. However, other factors may have been at play, given that more than half of the students incorrectly answered Question 3. This question was not included in the interview; instead, a new question (Question 5 in Figure 12.11) addressed similar issues.

Question 2 versus Questions 1 and 3. Question 2, which asked students to determine how long it would take to walk a total of 10 miles from the origin, may seem similar to Questions 1 and 3, which asked students, respectively, to determine how many miles Jose would walk from 3 to 5 hours and how long it would take for him to walk from 10 to 25 miles. From a mathematical perspective, the three questions ask about intervals. However, students may—and probably do—see them as different questions. While Questions 1 and 3 require students to consider and interrelate given intervals of distance and time, Question 2 allows students to bypass the concept of intervals and look at the single point corresponding to the distance of 10 miles and the time of 2 hours, thus arriving at the correct answer by identifying a point on the graph, rather than fully considering intervals.

Question 4. This question about how fast (in miles per hour) Jose walked, included only in the written assessment, was correctly answered by 59% of the students, an indication that the intervention successfully contributed to their understanding of linear functions as representing constant speeds.

Question 5. The additional question in the interview (see Figure 12.11, Question 5) probed students' understanding of the relationship between intervals in the two axes in more depth and was phrased as:

> Beginning at 4 hours, how much more time will Jose need to reach a total of 35 miles if he continues at the same pace? How do you know?

As already described by Caddle and Brizuela (2011), answers to this question by 21 students in the interview were as follows:

- Three students (14%) immediately answered 3 hours correctly.
- Seven students (33%) first answered 7 hours (total number of hours to reach 35 miles) and then provided the correct answer (3 hours), spontaneously or immediately after being asked, "But after 4 hours of walking, how many more?" They may have at first misinterpreted the question as referring to the time elapsed from the start of the walk to the 35-mile mark.
- Seven students (33%) tried to consider the interval starting at 4 hours but first gave the answer 2 or 4 hours, instead of 3. They all eventually arrived at the correct answer after some questioning from the interviewer.

Initial difficulties by these students were related to computational errors and to the challenge of coordinating the information on the two variables, a nontrivial task given that the question gave the start of the event as a point along the time axis and its end as a point along the distance axis.

- Four other students (19%) first answered 6 or 7 hours. Of these, one was not asked further about the interval from 4 hours to 35 miles, two were able to find the correct answer after substantial help, and one could not reach the correct answer. This last student remained unsure about whether points or intervals should be counted.

In summary, 80% of the 21 interviewed students correctly answered Question 5 in the interview.

Caddle and Brizuela (2011) analyzed in detail the strategies used by the same 21 students to answer Question 5, focusing on their use of multiplicative versus additive reasoning. They found use of scalar strategies (successive additive increment on one variable with parallel increments to the other) and functional strategies (use of multiplication based on the ratio between the two variables), as described by Vergnaud (1983), as well as mixed approaches.

Our analysis focused instead on each student's main strategy in terms of whether they considered the point-by-point correspondence across the two axes versus focusing mainly on the correspondence between the interval from 4 to 7 hours on the x-axis to the interval from 20 to 35 on the miles axis.

Five students focused on the correspondence between intervals, taking the previously established rate of 5 miles per hour and using it to find out that Jose needed 3 hours to travel 15 miles, as in the following example:

Interviewer: Beginning at 4 hours, how much more time will Jose need to reach a total of 35 miles if he continues walking at the same pace?
Student: Three more hours.
Interviewer: How do you know?
Student: 'Cause I did . . . I did, 3 times 15 and I know that every hour is . . . every hour he walks 5 miles so if you do 5 times 3 that would give you 3 hours and that would be 15 and then you do 20 plus 15 and then you do 35.

These five students considered the interval from 4 hours to 7 hours and/or the interval of distance from 20 miles to 35 miles, using the functional relationship between the two, without considering intermediate points.

Nine students, after determining that at 4 hours Jose had walked 20 miles, progressed 1 hour at a time through corresponding times and distances until they reached the desired point, thus moving from 4 hours and 20 miles to 5 hours and 25 miles, and so on, until 35 miles, as in the following example:

Student: Two more . . . wait . . . 3 more hours.

Interviewer: You think 3 more hours, why?

Student: Because when he has 20, he has another 5 hours and then another 5 and his last 5 hours will be at 35.

These nine students used the scalar approach and relied on specific points to structure a process of moving from one end point of the interval to another. At least in the explanations they provided, they did not seem to be using or operating on intervals along the axes. However, by working point-by-point, they could answer a question that referred to an event represented by intervals along the two axes.

The remaining seven students used an intermediary strategy, combining a point-wise approach with values for a short interval, thus jumping one of the steps in the process, as exemplified by student A's work:

Student: Because at 5 hours, he walked 25 miles, so like since every hour he walks 5 miles, I added 10 to this and got 35 [pointing to distance axis], and added 2 to this and got 7 [pointing to time axis].

This combined approach may constitute a step toward a full understanding of the relationship between intervals, as represented in the Cartesian space.

Achievements and Challenges in Understanding Intervals in Linear Function Graphs

Our results support the conclusions that (a) our early algebra intervention, with a strong focus on intervals along the number line and along the axes of the Cartesian space, helped children explore issues of co-variation and promoted understandings of relationships between intervals along the two axes in the Cartesian space; and (b) even though most elementary school children may prefer to focus on points, as opposed to intervals along the number line, they can use a point-by-point approach to determine how an interval on one axis corresponds to an interval on the other. Some of the students seem to have benefited from the intervention work on intervals and use strategies that combine consideration of points as well as intervals along the coordinate axes.

Most of the 5th-graders in this study inferred a value on one axis from its counterpart value on the other axis. For this relation, inferences could be validly made in either direction. They also determined the constant speed from the graph. Because the function was a special case of direct proportionality, a single coordinate (i.e., an ordered pair of values for distance and time at any instant) would suffice to compute the speed for any and all segments of the trip. A more telling test of students' understanding would require them to rely on changes in distance and time. This would be the case if the linear graph did not go through the origin or if the speed varied throughout the trip.

Regardless of the origin of their difficulties, by the end of 5th grade, intervention students outperformed control group students in written assessment items requiring consideration of intervals. In the interview, 47% of the students correctly answered a question requiring them to relate intervals along the two axes of a graph without help, and an additional 33% could do so with some help, leading to a total of 80% correct answers. Only four out of 21 students persisted in solely considering points while answering Question 1, and only one student was unsure about whether points or intervals should be counted when considering Question 5. At the end of the interview, 80% of the students reached a correct answer to Question 5. However, when answering this interview question, which required them to interrelate intervals along the two axes of a graph, only three students immediately focused on the correspondence of the interval of a total of 15 miles to a total of 3 hours. Other students, after recognizing that Question 5 asked them to start at 4 hours and to end at 35 miles, appropriately used the information in the Cartesian space, first relating points on the graph and on the axes and then using a scalar computational approach that considered the successive correspondence between number of hours and number of miles. Thus, the geometric representation as well as the arithmetic aspects of the task contributed to their solutions. From a starting point on the *x*-axis and an end point on the *y*-axis, they collected information held by individual points on the graphs, point-by-point, and combined them to arrive at the starting point in the *y*-axis and the end point on the *x*-axis. They then computed the number of miles and hours from beginning to end of the interval and arrived at the required quantification of the interval on the *x*-axis, even though they did not seem to operate on the interval as a whole.

DO FUNCTION GRAPHS BELONG IN ELEMENTARY SCHOOL?

Our findings show children's strengths and difficulties while they make sense of number lines and the Cartesian space representation. They support the conclusion that graphs of functions, usually absent from elementary school curricula in the United States, are within reach of students at this level. They also call for considering how learning and development interact in classroom settings, as children are given access to new representational tools. For some, it would seem unreasonable to ask children to deal with representations they are not yet ready to fully understand. For others, high expectations and new representations may help children perform beyond their initial resources. An example of this is the significantly better performance of the intervention students in the written assessment problem on the *K*-number line, where they used the number line as a tool to solve the problem. In addition, students participating in the intervention outperformed their control peers in interpreting and comparing graphs of linear functions, thus going beyond the usual exercises found in elementary school of determining point locations in the Cartesian plane. These results strongly suggest that the graphical representation of functions do belong in elementary school.

ACKNOWLEDGMENTS

This research was supported by the National Science Foundation grant # 0310171 to D. Carraher and A. Schliemann. We thank Bárbara M. Brizuela, Anne Goodrow, Susanna Lara-Roth, Darrell Earnest, Mara Martinez, Gabrielle Cayton, and Camille Burnett-Bradshaw for their contributions throughout the development of the project.

REFERENCES

Bass, H. (1998). Algebra with integrity and reality. Keynote address. In *Proceedings of the National Research Council Symposium on the nature and role of algebra in the K–14 curriculum* (pp. 9–15). Washington, DC: National Academy Press.

Bell, A., & Janvier, C. (1981). The interpretation of graphs representing situations. *For the Learning of Mathematics, 2,* 34–42.

Caddle, M. C., & Brizuela, B. M. (2011). Fifth graders' additive and multiplicative reasoning: Establishing connections across conceptual fields using a graph. *Journal of Mathematical Behavior, 30*(3), 224–234.

Carraher, D. W., & Schliemann, A. D. (2007). Early algebra and algebraic reasoning. In F. Lester (Ed.), *Handbook of research in mathematics education* (pp. 669–705). Greenwich, CT: Information Age Publishing.

Carraher, D. W., Schliemann, A. D., Brizuela, B. M., & Earnest, D. (2006). Arithmetic and algebra in early mathematics education. *Journal for Research in Mathematics Education, 37*(2), 87–115.

Carraher, D. W., Schliemann, A. D., & Schwartz, J. (2008). Early algebra is not the same as algebra early. In J. Kaput, D. W. Carraher, & M. Blanton (Eds.), *Algebra in the early grades* (pp. 235–272). Mahwah, NJ: Lawrence Erlbaum Associates.

Diezmann, C. M., & Lowrie, T. (2006). Primary students' knowledge of and errors on number lines. In P. Grootenboer, R. Zevenbergen, & M. Chinnappan (Eds.), *Proceedings of the 29th Annual Conference of the Mathematics Education Research Group of Australasia* (Vol. 1, pp. 171–178). Sydney, Australia: MERGA.

Freudenthal, H. (1983). *Didactical phenomenology of mathematical structures.* Dordrecht, Netherlands: Reidel Publishing Company.

Fuson, K. (1984). More complexities in subtraction. *Journal for Research in Mathematics Education, 15*(3), 214–225.

Gravemeijer, K. (1993). The empty number line as an alternative means of representation for addition and subtraction. In J. de Lange, I. Huntley, C. Keitel, & M. Niss (Eds.), *Innovation in mathematics education by modelling and applications* (pp. 141–159). Chichester, UK: Ellis Horwood.

Gravemeijer, K., & Stephan, M. (2002). Emergent models as an instructional design heuristic. In K. Gravemeijer, R. Lehrer, B. Van Oers, & L. Verschaffel (Eds.), *Symbolizing, modeling, and tool use in mathematics education* (pp. 145–170). Dordrecht, Netherlands: Kluwer.

Kaput, J. (1998). Transforming algebra from an engine of inequity to an engine of mathematical power by "algebrafying" the K–12 curriculum. In National Council of Teachers of Mathematics (Eds.), *The nature and role of algebra in the K–14 curriculum.* Washington, DC: National Academy Press.

Klein, A. S., Beishuizen, M., & Treffers, A. (1998). The empty number line in Dutch second grades: Realistic versus gradual program design. *Journal for Research in Mathematics Education, 29*, 443–465.

Kosslyn, S. M. (1989). Understanding charts and graphs. *Applied Cognitive Psychology, 3*, 185–226.

Leinhardt, G., Zaslavsky, O., & Stein, M.K. (1990). Functions, graphs, and graphing: Tasks, learning, and teaching. *Review of Educational Research, 60*(1), 1–64.

Peled, I., & Carraher, D. W. (2008). Signed numbers and algebraic thinking. In J. Kaput, D. W. Carraher, & M. Blanton (Eds.), *Algebra in the early grades* (pp. 303–328). Mahwah, NJ: Lawrence Erlbaum Associates.

Schliemann, A. D., & Carraher, D. W. (2002). The evolution of mathematical understanding: Everyday versus idealized reasoning. *Developmental Review, 22*(2), 242–266.

Schliemann, A. D., Carraher, D. W., & Brizuela, B. M. (2007). *Bringing out the algebraic character of arithmetic: From children's ideas to classroom practice.* Mahwah, NJ: Lawrence Erlbaum Associates.

Schwartz, J., & Yerushalmy, M. (1992). Getting students to function on and with algebra. In E. Dubinsky & G. Harel (Eds.), *The concept of function: Aspects of epistemology and pedagogy* (pp. 261–289). Washington, DC: Mathematical Association of America.

Treffers, A. (1991). Meeting innumeracy at primary school. *Educational Studies in Mathematics, 22*, 333–352.

Vergnaud, G. (1983). Multiplicative structures. In R. A. Lesh & M. Landau (Eds.), *Acquisition of mathematics concepts and processes* (pp. 127–174). New York: Academic Press.

Representational Re-Description as a Catalyst of Conceptual Change

Richard Lehrer and

Leona Schauble

Like the chapter authors, we have a long-standing interest in student-generated representations, both as instructional tools and as artifacts that can facilitate teachers' and researchers' understanding of student thinking and conceptual change. Students' struggles to represent their evolving thinking—through drawings, diagrams, maps, physical replicas, models, and mathematical descriptions—generate expressions of their ideas, and can also be catalysts for revising their thinking (Lehrer & Schauble, 2006). As children attempt to clarify their conceptions by externalizing them, they often perceive problems that were not apparent in their initially ambiguously specified thoughts. For example, in Chapter 10, Devin and Jamie began to explicitly consider that their musical instrument design might include both thick and thin rubber bands, and that some of the bands were short, whereas others were long. It is not apparent that these attributes were noted before the children attempted to arrange bands to produce different pitches; in fact, their early design work with the "instrument" seemed to focus primarily on the stability of the frame that held the bands. In this way, representing often instigates cycles of conceptual revision. Sometimes these cycles occur repeatedly, even over relatively brief periods of time. For example, initial sketches of a compost column show what one sees when one simply looks at it, but fail to reveal the larval forms hidden deep inside the compost. To solve this problem, a 6th-grader invents the "blow-up," a lens-shaped aperture that portrays a magnified version of the larva in question, anchored by a pointer that indexes its location inside the column. Over a few days, this solution is adopted by the class as a conventional form for portraying things that are deemed important, but are not visible to casual inspection. Because the "blow-up" portrays a magnified version of its subject, students begin to inspect the larval forms more carefully and identify morphological differences among the larval kinds that they did not originally notice (Lehrer & Schauble, 2012).

The widespread appropriation of the "blow-up" illustrates another instructional advantage of representations, namely, that they usually have intended audiences, who provide an additional source of feedback. A 2nd-grader tries to explain to a peer how she manipulated a quilt "core square" to make a complex quilt design. Her first explanation relies heavily on gestures, but the gestures are not stable over time and are approximate, at best. As a result, her classmate is unable to replicate the design. Over successive attempts to clarify instructions for making a "quilt design," the class first agrees on a repertoire of transformations and terms to describe them (such as flips and turns—including full turns, half turns, and quarter turns, all further specified by turn direction), and eventually, on abbreviations that are incorporated into replicable procedures (Jacobson & Lehrer, 2000).

Student representations typically are met with objections, ratifications, or alternative proposals from peers that may be ignored, adopted, or partially assimilated. Hence, the generation of representations is, in our view, only a first step. We also favor instructional contexts in which students use both their own representations and those developed by others to solve new problems. Attempts to do so frequently bring students' thinking into productive contact and serve as the engine for new perspectives. This practice also fosters explicit consideration of the tradeoffs entailed by alternative representational choices. When students repeatedly assess what different representational devices "show and hide," they are building what diSessa and Sherin (2000) call "meta-representational competence." Putting representations into contact with each other is critical for developing this form of competence. For example, one pair of students featured in Chapter 10 created a physical representation of pitch by having rubber bands stand in for features of a vibrating string, whereas a second pair created a representation governed by physical resemblance to real instruments. Although Wendell does not tell us whether the teacher juxtaposed these two contrasting representations of "an instrument" for the class at large, our guess is that doing so would be a valuable pedagogical move. These differences in perspective about the role and function of representation were consequential, and it would be important for future instructional design to mine this variability for its opportunities to help students understand relations between representational and performative idioms of expression (Pickering, 1995).

We are particularly interested in representations, along with the contexts that draw on representational use, that encourage continual "upping the ante," so that initially accessible representations, like drawings, are gradually connected to increasingly more powerful (but cognitively challenging) forms, such as labeled diagrams, magnifications, and Cartesian systems. Third-graders who were studying the ecological succession of plants in a "wild backyard" began by adopting a "personal plant," which they first drew in the outdoors and then studied in the classroom. As the class members began to compare the features of their plants, drawings gave way to labeled diagrams, so that students could more readily contrast plant structures across species. Many of the plants were dug up to expose their root systems (and immediately began to die). Fall turned toward winter, and the diagrams needed now to express observed changes over time. As the 3rd-graders began to wonder "how all those plants got to the backyard," the teacher introduced Videoflex™ magnifications of seed parts so that the children could observe and, subsequently, test hypothesized forms of seed travel. Speculating that different plants might have different "needs," students constructed a scale map of the backyard and located each plant, to more precisely characterize the conditions of moisture and shade where each species was most abundant. In this case, successive refinements and representational re-descriptions both were inspired by and, in turn, inspired changes in the conceptual space that the plants occupied for these 3rd-graders (Manz, 2012).

Of course, general enthusiasm for representational approaches should not be mistaken for a belief that they can be an all-purpose panacea. From a pedagogical perspective, educators should ask themselves questions like: What work is the representation intended to accomplish? Am I asking merely for an artistic expression, or does this representation fulfill a role in helping students remember, elaborate, explain, even find out? For example, did the 1st-graders in Chapter 11 really need to draw the components for their tool to fetch the keys from the Plexiglass box? Given the limited choices of parts, as well as the students' capability to try out intermediate constructions, we wondered whether the children perceived the drawings as serving as incubators of design or merely as something that had to be accomplished before they could construct devices. Certainly, previous research establishes that even much younger children do not need memory aids to select the right component from choices to complete a tool for accomplishing a simple mechanical purpose (Bullock, Gelman, & Baillargeon, 1982). Similarly, does the representation have a lifespan in the classroom beyond its initial construction? Is anyone, either the inventor or preferably the entire class (and even individuals beyond the class), going to use this representation to tackle new problems in the future? In general, we try to avoid asking students to generate one-time solutions, representations that go into a drawer and are never again used or built upon. What that tends to communicate is that drawing, diagrams, graphs, and models are just another kind of school task, rather than cognitive tools of genuine utility.

Similarly, educators should note that mastering representational forms has costs, as well as benefits. Our enthusiasm about their potential should not blind us to the costs. Sometimes those costs are extensive, but, on reflection, they are clearly worth the time. These are usually representational systems, like mathematics, that have an extended life through a student's educational years and beyond.

Good examples are the number line and the Cartesian plane featured in Chapter 12. The 3 years devoted to understanding the implications of these representations are warranted by the importance and ubiquity of these symbolic systems in mathematics and science. Here, careful instructional design positioned students to see the representational utility of aligning measure spaces along a perpendicular instead of a parallel. Moreover, the problems posed to students blended metaphors of quantities as measures and as points along pathways of motion (Lakoff & Nuñez, 2000). However, the comparatively low levels of absolute performance on some of the critical measures of student understanding suggest that learning about these representational systems presents formidable challenges to instructional design. On the other hand, the learning costs of other potentially promising representational systems can be extensive and the long-term gains more questionable. A common example is computational technologies, whose adoption and use suffer from the generally impoverished infrastructure for supporting them that exists in many of our schools. Many new technologies, including some very exciting dynamic representations, are potentially of great learning benefit, but in some cases, their adoption and continued use across years of instruction are difficult to predict with confidence. This makes it more difficult to assess the likely payoff for investing in the time and energy to learn them. We certainly are not arguing against them, but we do acknowledge the need for a clearheaded assessment of the stakes involved.

Finally, from a research perspective, we maintain some skepticism about presuming an unproblematic mapping between the representations and artifacts that children produce and their own internal mental representations, or conceptions. Although representations and models can help make children's thinking more visible, perhaps they should not be accepted too literally as externalization of what is in the mind. This caution is one of the reasons that Latour (1990) designated such representations as inscriptions. The artifacts that children construct are guided by their ideas but are also constrained by the materials that are available and children's strengths and limitations in assembling them, as Portsmore points out in Chapter 11. Often, the model a child makes is not the model he or she intended, not because of some change in plan or insight, but because the materials resisted the original intentions. A child intends to make a LEGO® Mindstorms™ car with a complicated gear system but is unable to mount all those gears on the platform that he constructs. This intractability of materials holds true even with artifacts as seemingly accessible as drawings. For example, children who attempted to draw "two houses, one far and one near" struggled to figure out how to draw two objects on the same page when one partially occluded the other (Braine, Schauble, Kugelmass, & Fell, 1993). This is an interesting cognitive problem: How do you show something that you cannot see? There are, of course, representational conventions for mastering it, but most young children do not know them. Moreover, the problem children solve when they are constructing an artifact may or may not be the problem intended by the adult who posed the task. To the chagrin of educators, it is well known that many design problems can be solved by trial and error, or by some new, unforeseen solution, rather than by employing the scientific principle that the context was designed to exemplify. It may be tempting to conclude that the children in Chapter 10 understood tension because they successfully designed a musical instrument made of rubber bands set at different tensions. On the other hand, the evidence equally supports the alternative possibility that children only noticed that some rubber bands were thick and some were thin; moreover, that they could be wound once or twice around the supporting frame. In this case, one can legitimately wonder what serves as convincing evidence that the children "had the idea" of tension.

In sum, we strongly support the emphasis on representational forms that is displayed in all of these chapters. We applaud the attempts to go beyond simply asking children to draw, symbolize, or make, and to find out with more precision what they think they are doing when they use these representational devices, especially when representations, materials, and performances are intertwined, as they are in these chapters. We appreciate the opportunity to reinvigorate dialogue about the affordances and challenges of representations and symbolic systems, both for supporting and for studying children's thinking.

REFERENCES

Braine, L., Schauble, L., Kugelmass, S., & Fell, A. (1993). The representation of depth by children: Spatial strategies and lateral biases. *Developmental Psychology, 29*(3), 466–479.

Bullock, M., Gelman, R., & Baillargeon, R. (1982). The development of causal reasoning. In W.J. Friedman (Ed.), *The developmental psychology of time* (pp. 209–254). New York: Academic Press.

diSessa, A. A., & Sherin, B. L. (2000). *The Journal of Mathematical Behavior, 19*(4), 385–398.

Jacobson, C., & Lehrer, R. (2000). Teacher appropriation and student learning of geometry through design. *Journal for Research in Mathematics Education, 31*, 71–88.

Lakoff, G., & Nuñez, R. E. (2000). *Where mathematics comes from. How the embodied mind brings mathematics into being.* New York: Basic Books.

Latour, B. (1990). Drawing things together. In M. Lynch & S. Woolgar (Eds.), *Representation in scientific practice* (pp. 19–68). Cambridge, MA: MIT Press.

Lehrer, R., & Schauble, L. (2006). Cultivating model-based reasoning in science education. In R. K. Sawyer's (Ed.), *Cambridge handbook of the learning sciences* (pp. 371–387). Cambridge, UK: Cambridge University Press.

Lehrer, R., & Schauble, L. (2012). Supporting inquiry about the foundations of evolutionary thinking in the elementary grades. In S. Carver & J. Shrager (Eds.), *The journey from child to scientist: Integrating cognitive development and the education sciences* (pp. 171–206). Washington, DC: American Psychological Association.

Manz, E. (forthcoming). The co-development of modeling practice and ecological knowledge. *Science Education.*

Pickering, A. (1995). *The mangle of practice: Time, agency and science.* Chicago, IL: University of Chicago Press.

Formulating Measures

Toward Modeling in the K–12 STEM Curriculum

Judah L. Schwartz

ON MEASURES AND MODELS

In science and engineering education, all too often problems involve *selecting* an already formulated model that describes/explains the phenomena in question followed by selecting the "correct" relation among relevant aspects of a situation and quantitatively evaluating it. If, however, science and engineering education is about teaching people how to build and use ever more effective models of natural phenomena, should we not focus this education on formulating the relations and the elements that are related by these models? We do focus on relations embodied in our models such as how the distance traveled by a falling body depends on its initial velocity and the height from which it is released, or how energy is transformed from energy of motion into energy of configuration (and back) in systems that oscillate. The quantities that are the elements that these relationships associate—kinetic energy, potential energy, velocity, acceleration—are all measures that we tell our students are worth attending to. However, our students have no hand in formulating these measures that are the elements of the models we build and use.

It is important to stress that measures are not models—rather, they are the *elements* of models. In the sciences, models describe how measures are related to one another, and often, how they vary in time and/or how they vary in space.

This chapter is about the role of formulating measures in the building of models. It is not a traditional report on empirical studies contrasting one group's performance with another. Rather, it is an analysis of the structure of models and the relationship between measures and models, the product of careful reflection about several decades of teaching science and engineering students and teachers.

It is intended to be something between a plea and a manifesto—calling for more detailed understanding of how the elements of models are formulated and assembled into the tools of scientific thought.

How does this all relate to the theme of this volume, which deals with representations in science and mathematics education? As we shall see, the relationship between representations and the formulation of measures is clear and direct. Measures are, in many respects, the quantitative instantiations of the elements of representations.

It is centrally important to the practice of many natural and social science disciplines to form measures of quantities of interest. By *measure*, I mean a construct that we devise to quantify the amount or degree of a property or attribute of an object or a situation of interest. Indeed, the formulation of measures is an important first step in the act of modeling—with the measure itself serving as the heart of a structural model, or the time dependence of the measure as the core of a functional model (e.g., Schwartz, 2007).

Geographers (e.g., Wong, 2005) studying spatial segregation patterns, computer scientists studying the evolution of biological complexity[1], biochemists (e.g., Heymans & Singh, 2003) trying to establish the similarity of metabolic pathways, reading specialists[2] studying dyslexia among children, and public health physicians (e.g., Steptoe & Feldman, 2001) studying stress in neighborhoods are but a few examples of measures devised by scientists working in diverse disciplines.

The essence of all such measures, and indeed any measure, is the fact that it can be ordered. Sometimes we seek to order quantities that are directly accessible to the senses, such as heights and weights. The rank ordering of such quantities is relatively straightforward, although it may require the introduction of instruments to refine or extend our senses. Sometimes, however, we seek to order quantities that are clearly dependent on several, or even many, contributing factors such as cost of living or the efficiency of an automobile.

It is important when ordering such complex quantities that the bases for the orderings be made explicit and public. This may permit one to say that in ordering A, B, and C whether B is closer to A than it is to C and if so, by how much. When this happens, it is fair to say that we have devised a measure of the scale along which A, B, and C are arrayed.

This chapter is a plea for a new kind of task in the K–12 STEM curriculum—one that asks students themselves to formulate measures rather than simply using the measures provided them—as a step in the direction of understanding how models are made. Needless to say, I do not expect students to invent constructs such as kinetic energy or momentum—I seek only to have them ponder the situation they are trying to understand hard enough to recognize, for instance, that a measure of momentum must have a time rate of change, which is dimensionally the same as a force.

WHAT ARE MEASURES?

Measures are either directly observed continuous or discrete quantities; or, measures are mathematical constructs computed from observed continuous or discrete quantities. We need to sharpen the distinction between assigning magnitude to discrete quantities and assigning magnitude to continuous quantities. In attaching a magnitude to a count noun, we have to make a judgment that takes all the attributes of the referent object into account and that results in a single yes-or-no decision as to whether the object we are considering does or does not belong to the class of things to which we are assigning number. It goes without saying that the only sort of number that can result from the act of counting is an integer.

However, in the case of mass (or continuous) nouns, we can introduce adjectival quantity only after we have singled out the attribute of the referent object that we wish to quantify. We say, for example, 300 grams of clay. In so doing, we ignore the shape of the clay and its color and focus solely on its weight (more properly its mass, in this case). Because of these considerations, it is clear that the result of assigning size to some attribute of a mass noun, an act we normally call *measuring*, necessarily results in a nonnegative rational number.[3]

Perceptually Available (Simple) Measures

Humans are able to assign at least relative size to certain sorts of measures, which can then in turn be combined into composite measures such as quotient and product measures, which are the main subject of this chapter. We have to be careful how we say this: We can order counts if they are not confounded with differing spatial attributes, as is the case in many conservation experiments; we can order continuous quantities such as lengths, times, and weights if they are not beyond the resolving power of our perceptual apparatus. These measures include counts, lengths, areas, times, weights, and speed.

Composite Measures—*Qualitative* Relations Among Simple Measures

We can form combinations of perceptually available measures in order to generate more complex measures that may be more informative in more complex situations. Clearly, the simplest form of combination is one that brings together two simple measures. We consider first a form of combination that leads to direct variation. We refer to such combinations as product measures. A product measure is a composite measure C that depends on two component measures A and B such that:

if A ↑ then C ↑—*more* A → *more* C
if B ↑ then C ↑—*more* B → *more* C

Here is a situation that leads to the formulation of a product measure: Assume we have a collection of objects (say, pickup truck, fast-pitched baseball, car, motorcycle, bicycle, and so on) that are moving. Most people would be willing to agree that there is some property of a moving pickup truck that is usually "larger" than that of a moving bicycle. Suppose, for the moment, that we call this property "oomph."

Is there some way we can pin down what such a property called "oomph" might be? More to the point, is there some way we can say whether the "oomph" of some other moving object is closer to the "oomph" of the pickup truck than it is to the "oomph" of the bicycle? And if so, how much closer?

More generally, if we have a bunch of different moving objects, can we find a way to order them according to the "oomph" property—from the one with the "largest" "oomph" to the one with the "smallest"? First thoughts would suggest that the relevant properties of the moving objects that we want to pay attention to are how heavy they are and how fast they are moving. Each of the moving objects has both a mass and a speed.

Assuming we have proper instruments for measuring speed and mass, we could measure both the speed and the mass of each of the moving objects. If we do so, we will quickly discover that our mystery property, "oomph," cannot be mass and it cannot be speed. A baseball, which is clearly less massive than a car, can be much harder to stop than a slowly rolling car. By the same token, even a very slowly rolling locomotive can be much harder to stop than a fast-moving car. Clearly, any measure of this "oomph" property has to take *both* speed and mass into account.

Let us put on hold for the moment making such a measure quantitative and turn to its logical complement—a form of combination that leads to inverse variation. We refer to such combinations as quotient measures. A quotient measure is a composite measure C that depends on two component measures A and B such that:

if A ↑ then C ↑—*more A* → *more C*
if B ↑ then C ↓—*more B* → *less C*

Here is a situation that leads to the formulation of a quotient measure: If we have a collection of pieces of aluminum (say, an aluminum saucepan, a door hinge, a machine screw), a collection of pieces of pine wood (say, a cutting board, a small jewelry box, a chess piece), and a collection of pieces of polystyrene (say, several different food containers), most people would be willing to agree that there is some property of aluminum that is "larger" than that of pine. Suppose, for the moment, we call this property "oomph."

Is there some way we can pin down what such a property called "oomph" might be? More to the point, is there some way we can say whether the "oomph" of some third material like polystyrene is closer to the "oomph" of aluminum than it is to the "oomph" of pine? And if so, how much closer?

More generally, if we have a bunch of different materials, can we find a way to order them according to the "oomph" property—from the one with the "largest" to the one with the "smallest"? First thoughts would suggest that the relevant property of the aluminum pieces, the pine pieces, and the polystyrene pieces that we want to pay attention to is how "big" they are. But by "big" we might mean their physical size (volume—the amount of space they take up) or we might mean their weight (or their mass). Clearly, each piece of aluminum, pine, and polystyrene has *both* a volume and a mass.

While product measures and quotient measures do not exhaust the gamut of measure structures that are used in scientific models, they are the simplest measures that can be formed from the perceptually available measures of lengths, times, weights, and so on.

Composite Measures— ## *Quantitative* Relations Among Simple Measures

In many situations we go beyond the qualitative variation considered in the previous section. Consider, for example, a product measure such as momentum, by returning to the qualitative attempt at formulating a measure that captures a direct variation—the problem of stopping a moving object: In order to formulate a measure of "oomph," we could try to "add" measures of speed and mass. However, doing so leads to the unfortunate result that a body could have "oomph" even if it were standing still. Also, this kind of measure of "oomph" leads to the conclusion that even a particle with no mass can have "oomph."

A more successful measure of "oomph" would be a product of mass and speed. This sort of measure avoids the difficulties that the previous proposed measure had. With this measure, a moving object has no "oomph" if it is standing still. Moreover, even at a fixed speed, the smaller the mass of the moving object, the smaller its "oomph." This measure of how difficult it might be to stop a moving object is normally not called "oomph" but is called *momentum.*[4]

Now let's turn to building a quantitative product measure, one that is well known to be difficult for middle school and even upper elementary, students (Finesilver, 2009). The measure in question in this example is "Size of Available Menu" (see Figure 13.1).

The most common sorts of difficulties middle school students encounter with this kind of problem are twofold. The first sort of difficulty is a "modeling" difficulty—rather than construct a measure, they simply enumerate as many of the possible sandwiches as they can. Needless to say, this leads to omissions and counting difficulties. The second sort of difficulty comes from the formulation of an inappropriate (and dimensionally incorrect) measure, namely the sum of the number of kinds of cheese, meat, and bread. In the present case, this leads the students to be unable to choose whether shop A or shop B offers the larger selection.

Figure 13.1. "Size of Available Menu" Example

Two Sandwich Shops

Shop A offers 3 kinds of cheese, 6 kinds of meat, and 3 kinds of bread.

Shop B offers 4 kinds of cheese, 4 kinds of meat, and 4 kinds of bread.

Which shop has the larger offering? How much larger?

Design a menu offering that is:

larger than the smaller of these two &

smaller than the larger of these two.

Formulating product measures can be a problem in many different contexts and may demand different approaches depending on context. Indeed, sometimes it is difficult to say what product measure is best suited for the analysis of a given situation. Consider the following example, appropriate for secondary students, of a measure of efficiency for different aircraft types taking into account the Number of passengers that can be carried, Average speed as miles per hour [mi/hr], Average range [mi], Average fuel consumption as gallons per hour [gallons/hr], and Operating cost as dollars per hour [$/hr].

Product measures are found throughout the universe of models. Typically, they are Cartesian Products—here are some examples: area, volume, work, momentum, passenger-miles, kilowatt-hours, and flux. Such product measures are almost always extensive quantities and are rarely (if ever) dimensionless.

Let us now return to the attempt at formulating a measure that captures an inverse variation—specifically, the problem of quantitatively characterizing a material rather than an object made of that material. Assuming we have the proper instruments for measuring volume and mass, we could measure both the volume and the mass for each of the pieces of aluminum, pine, and polystyrene. If we do so, we will quickly discover that our mystery property, "oomph," cannot be mass and it cannot be volume. We can find some pieces of pine with more mass than pieces of aluminum and some pieces of pine with less mass than pieces of aluminum. Table 13.1 displays data from the measurement of masses and volumes and a scatter plot of those data typically found in chemistry and physics handbooks.

A scatter plot of these data looks like that shown in Figure 13.2.

Normally, we present the concept of density sometime in middle school as the ratio of mass to volume. If, however, we proceed from the data, we see that doing that constitutes a leap! We have a finite number of data points—the assertion that

density = mass/volume

Table 13.1. Masses and Volumes for Objects Made of Three Different Materials

MATERIAL 1			MATERIAL 2			MATERIAL 3		
Aluminum			Pine			Polystyrene		
volume (cc)	*mass (gm)*	*gm/ cc*	*volume (cc)*	*mass (gm)*	*gm/ cc*	*volume (cc)*	*mass (gm)*	*gm/ cc*
0	0	0	0	0	0	0	0	0
120	324	2.7	15	8.3	0.55	41	43.1	1.05
73	197	2.7	62	34.1	0.55	103	108.2	1.05
31	83.7	2.7	127	69.9	0.55	79	83	1.05
110	297	2.7	90	49.5	0.55	67	70.4	1.05

Figure 13.2. Masses and Volumes for Objects Made of Three Different Materials

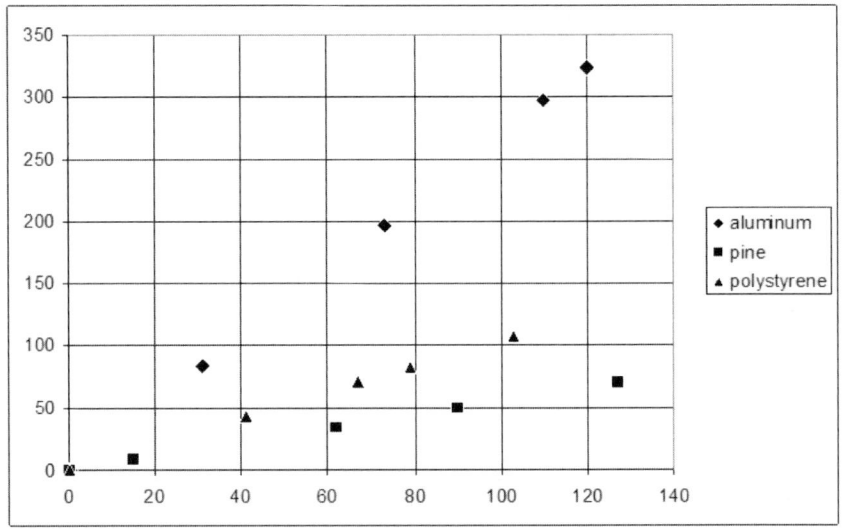

for all nonnegative values of mass and volume is an assertion of a model. It is a model allowing us to predict the magnitude of the mass of a piece of material whose volume we have measured but whose mass we have not measured. Similarly, the model allows us to predict the value of the volume of a piece of material whose mass we have measured but whose volume we have not measured. Quotient measures such as density are almost always intensive quantities and are sometimes dimensionless.[5] There is a caveat that must be expressed here about this model of density, which is fundamentally a macroscopic model of density. It depends on the assumption that seems to work well in our macroscopic world (i.e., that when two masses are put together, their combined mass is the sum of their masses, and that when two volumes are put together, their combined volume is the sum of their volumes). This assumption may not be valid on either atomic or galactic spatial scales.

Perceptually Based Quotient Measures

As a way of ramping up to the introduction of measure-formulating tasks into the curriculum, I present here several such tasks that are largely based on visual perception and that require little or no external or prior knowledge to begin the task. In fact, the Balanced Assessment in Mathematics project (2007) from which these tasks have been drawn has used the opening sections of the tasks in 1st- and 2nd-grade classrooms. The full range of the tasks has been widely used at the secondary level and the tasks and their extensions have been used in graduate courses in mathematics education. Because these tasks tend not to make demands on prior or external knowledge, and because they do not involve previously encountered formally defined measures, they present a nice opportunity to have students devise measures and to discuss their relative merits (see Figure 13.3).

Here are some of the approaches to this problem taken by middle school and upper elementary students. The approaches described are composites of observations in different classrooms and discussions led by different teachers.

Figure 13.3. "Square-Ness" Problem

Below is a collection of rectangles.

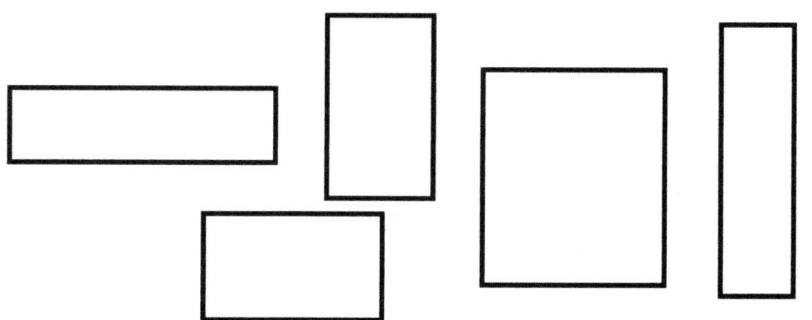

1. Which of the rectangles is the "squarest"?

2. Arrange the rectangles in order of "square-ness" from most to least square.

3. Devise a measure of "square-ness," expressed algebraically that allows you to order any collection of rectangles in order of "square-ness."

4. Devise a second measure of "square-ness" and discuss the advantages and disadvantages of each of your measures.

Approaching the Problem with a Subtractive Comparison of Lengths. Let us denote the length of a horizontal leg of a rectangle by H and a vertical leg by V. A first attempt at a measure of "square-ness" might then be $H - V$. Thus, a rectangle with horizontal sides of 6 cm and vertical sides of 2 cm has a "square-ness" of 4 cm, while a rectangle with horizontal sides of 2 cm and vertical sides of 6 cm has a "square-ness" of −4 cm.

Upon hearing this approach, a teacher might intervene in the following way: Is it desirable to have congruent (i.e., "same shape—same size") rectangles have different measures of "square-ness?" Probably not. How, then, might we revise our measure to fix this?

Suppose we take as our measure of "square-ness" $H - V$. Some of the older middle school children that we have worked with on the Balanced Assessment project actually suggested the use of the absolute value function by saying, "Just take the difference between H and V." Clearly, our problem with congruent or "same shape—same size" rectangles is now repaired.

The teacher might press further: What about two rectangles that are similar (i.e., "same shape—not same size")? Consider the following problem: Given two rectangles, one with horizontal sides of 6 cm and vertical sides of 2 cm, and the other with horizontal sides of 18 cm and vertical sides of 6 cm. The first rectangle has a "square-ness" measure of 4 cm and the second has a "square-ness" measure of 12 cm. Is it desirable to have similar ("same shape—not same size") rectangles have different measures of "square-ness"? How might we revise our measure in order to achieve this goal? With substantial prodding, the class decides to divide the measure of "square-ness" by the perimeter (actually, the half perimeter) of the rectangle. The justification is that bigger rectangles have bigger perimeters, and thus, there is some chance that this problem can be repaired. Suppose we take the quantity $\frac{|H-V|}{H+V}$ as our measure of "square-ness." This measure repairs the problem of similar rectangles not having the same measure of "square-ness." Note that this measure of "square-ness" is dimensionless and has values that do not depend on the length unit used to measure H and V.

Are we done now? To some extent, this is a matter of taste. This measure of "square-ness" has the peculiar property that the "square-ness" measure of a square is 0. The "square-ness" measure of a very wide and low rectangle approaches 1. Although it is less easy to see, it is also true that the "square-ness" measure of a very narrow and tall rectangle approaches 1. Wouldn't it be nice to have the measure have the property that the "square-ness" of a square is 1 and that of a very elongated rectangle approach 0? We could define a new measure $1 - \frac{|H-V|}{H+V}$. This measure now has the features we would like our measure to have.

Are we done? Not quite. It looks like this measure depends on two variables, H and V. Actually, it is possible to rewrite the form of this measure so that it depends on only one variable. For instance, define the ratio $r = \frac{V}{H}$. Divide numerator and denominator in the expression for "square-ness" by H. Then, with the aid of this definition, it is possible to rewrite our measure of "square-ness" as $1 - \frac{|1-r|}{1+r}$.

Another Subtractive Comparison of Lengths Approach. Another
approach taken by many students is to attend not to the horizontal and vertical
orientations but rather to the longer and shorter sides of the rectangles. Suppose
we designate the longer side of the rectangle by L and the shorter side by S. The
subtractive comparison now leads to L − S as a measure of "square-ness." The
problem with the sign of the measure is no longer present, but the problem of
similar rectangles having different measures of "square-ness" remains. Once again,
this problem is resolved by dividing by the (half) perimeter, leading to $\frac{L-S}{L+S}$ as a
measure of "square-ness."

Approaching the Problem with a Ratio Comparison of Length.
There is another line of thinking about this problem. Using the same notation of
H and V for the horizontal and vertical sides of the rectangle, some students take
the ratio $\frac{H}{V}$ directly as a measure of "square-ness." This measure has the property
that it does not depend on the units in which the lengths of the sides are measured. It does, however, suffer from the following problem: Consider a rectangle
with horizontal sides of 6 cm and vertical sides of 3 cm. According to this measure,
the rectangle has "square-ness" 2 while a rectangle with horizontal sides of 3 cm
and vertical sides 6 cm has "square-ness" $\frac{1}{2}$.

Once again, middle school students are not happy with the measure of
"square-ness" being dependent on the orientation of the rectangle. Here is a pos-
sible resolution of this difficulty. Let us revise the measure so that it is $\frac{H}{V}+\frac{V}{H}$. The
rectangle in question then has "square-ness" 2.5 in both orientations. This measure
of "square-ness" has the peculiar property that the "square-ness" of a square is 2.
We can fix this by revising our measure so that it is $\frac{H/V + V/H}{2}$. Is this a satisfactory
measure? There are those who would argue that this measure is not yet satisfactory because the "square-ness" measure of very elongated rectangles grows without
limit. In addition, although the "square-ness" measure of the square is 1 under this
measure, the "square-ness" measure of any rectangle that is not a square is larger
than 1. Wouldn't it be nicer if the "square-ness" measure of non-square rectangles
were smaller than the "square-ness" measure of squares?

Another Ratio Comparison of Lengths Approach. Middle school
students who focused on longer and shorter sides rather than horizontal and
vertical sides eventually arrived at either S/L or L/S as a measure of "square-
ness." The measure S/L generally seemed to be preferred because all rectangles
had "square-ness" measures that lay between a "square-ness" of 0 and a "square-
ness" of 1.

Approaching the Problem Through Area Comparison. Another
approach used by a few students was to compare the area of a square formed by
the longer side to that of a square formed by the shorter side. There are other
ways of measuring "square-ness" based on a comparison of areas. For example,

suppose one considers the area of the square whose side is L and the area of the square whose side is S. Then a possible measure of "square-ness" is $L^2 - S^2$. This measure does not have the difficulty that congruent ("same shape—same size") rectangles in different orientations have different measures of "square-ness." It does, however, suffer from the problem of geometrically similar ("same shape—not same size") rectangles having different values of their "square-ness" measures. In the same spirit of revision as we employed earlier, we can define a new measure $\frac{L^2 - S^2}{L^2 + S^2}$. This expression is also the ratio of two areas. I invite the reader to consider possible extensions of the square-ness task to parallelograms and to parallelopipeds.

The story doesn't end there. In fact, it probably doesn't end at all. The moral of the story is that humans make mathematics, and that the problems we pose to our students ought to offer them the opportunity to do the same.

Some other interesting tasks that require little or no prior information and rely on the formal definition of measure on perceptually available images include sharpness of turns in a road, smoothness of spheres (tennis ball, golf ball, Earth, and so on), disc-ness (dime, tuna fish can, piece of uncooked spaghetti, and so forth).

As a further example of tasks that depend on formally defining measures based on visually perceived images, we explored the issue of densities of all sorts in the Balanced Assessment project. We did so because of the large research literature on children's understanding of (mass) density (e.g., Kloos, Fisher, & Van Orden, 2010) as well as our own work at the Fulcrum Institute (2010) on teachers' ability to generalize the concept of density beyond mass/volume to linear, areal, and volume densities to such attributes as populations (e.g., populations/area), energy (e.g., energy/volume), and counts (e.g., automobiles/length of road). The original task was presented to middle school students in several Boston-area schools in the form shown in Figure 13.4.

Following the original presentation of the task, we presented a modified version that was contextually situated in a familiar situation, i.e., photographs of beaches with varying densities of sunbathers, as a counterpoint to the abstract nature of the original form of the task. Middle school students' responses to the original task (see Figure 13.4) were interesting. Most of the children measured all the dot-dot distances for each case. Each group in Figure 13.4 has 8 dots and thus there were (8 x 7)/2 distances for each group. This necessitated a systematic approach to measuring and recording the distancing—a task which itself took some time and effort. There was, however, a good deal of debate about what to do with the collection of distances. Some children felt that the average dot-dot distance was a good measure. A few felt that the ratio of the largest dot-dot distance to the smallest dot-dot distance was a good measure. There was a lot of discussion about what the collection of dot-dot distances would look like in the case of a regular array, such as a crystal or tile floor.

Perhaps the most interesting proposed measure came from one of the 6th-grade students who said (paraphrased), "Imagine each set of dots is a set of nails

Figure 13.4. Densities of Different Kinds

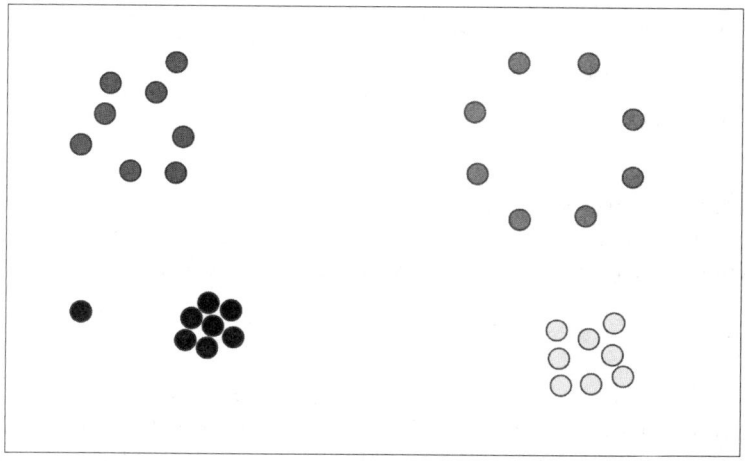

Which of these collections of dots is the most "crowded"? The least "crowded"?

Can you define a measure of "crowded-ness" that allows you to order any collection of a [fixed] number of dots in order of "crowded-ness" from least crowded to most crowded?

---

on a Geoboard. Put a rubber band around all the nails in the set and measure the area enclosed by the rubber band. Then a good measure of crowded-ness is the number of nails divided by the area enclosed by the rubber band." The reader may recognize that this student invented the idea of a "convex hull"—the smallest convex polygon to enclose all the members of the set.

Densities are examples of quotient measures and are thus intensive quantities with all the understanding difficulties attending such measures. Generalizations of density that we found science teachers having difficulty with include *linear* densities as in the case of traffic flow, electric charge density and energy density in energy storage media such as batteries or capacitors, as well as energy density in propagating electromagnetic waves (Fulcrum Institute, 2010). Among the questions that arose in discussion with middle school science teachers was the following: "If energy coming from the sun arrives at the Earth, then there is energy at the sun and there is energy at the Earth. Is there energy in the space in between? How much energy? Is there more energy in a larger volume of space? And so on."

Additive Measures

Are there measures that are more complex than product measures and quotient measures that play a role in K–12 education? Here is an example: The cost of renting a car can depend on both the number of days the car is rented as well as the number of miles driven (see Figure 13.5).

Figure 13.5. Renting a Car Problem

Total cost {in \$} = d {in \$/day} x D {in days} + m {in \$/mile} x M {in miles}
Total cost is an additive measure = Time Cost + Distance Cost
Time Cost is a product measure
Distance Cost is a product measure

Additive measures are more complex than quotient measures and product measures. Indeed, they can be more complex than simple combinations of pair of product or quotient measures. Additive measures are the form:

$$c_1 A_1 + c_2 A_2 + \ldots + c_n A_n$$

where the set $\{A_i\}$ denotes the set of attributes that are deemed to be important for the composition of the measure. The set of coefficients $\{c_j\}$ is used to make the measure dimensionally coherent, and they also express the relative importance that the person formulating the measure attaches to the different attributes in composing the measure.

In general, measures in the social sciences have this sort of structure. Probably the best-known example is Gross National Product (GNP), a combination of a large number of indicators with suitable coefficients so that the combined measure is in dollars.[6]

Is the Measure "Good Enough"?

Choosing a measure to characterize the state of a system is an exercise in deciding what features of the system one wants to include in one's characterization and what features are sufficiently unimportant for the purpose at hand that they may be neglected, at least at first. For example, a measure of density may not, at first, contain any reference to temperature. Thus, the process of formulating a measure as a first step on the road to modeling may be tentative and provisional. Ultimately, the question of "is the measure good enough?" can only be answered by listening to the fugue of the predictions of the model formulated with this measure and the purpose for which the model has been formulated.

FROM MEASURES TO MODELS

Measures That Are Models and Measures That Are Not Models

Are measures models? That depends. In my view, an essential property of a model is its ability to predict the value of a measure in an *as yet untried instance*. In the case of density discussed above (see Table 13.1), we designed a measure (and called it density) based on data collected from a variety of objects that were fashioned from a variety of materials. Can we *predict* the value of the density measure for an object whose mass and volume we have not previously measured? Under limited circumstances, the answer is yes. If the material is one with which we have worked before, then using our rule for how mass and volume are related for the material in question:

- if we measure the mass, we can *predict* the volume
- if we measure the volume, we can *predict* the mass
- if we measure the mass and the volume, we can *calculate* the ratio of mass to volume (it has to be stressed that for a finite number of data points the inferred linear mathematical relationship between mass and volume is *not* unique; the assumption of a linear relationship is part of our assertion about the way nature behaves in the *as yet untried instances*)

On the other hand, if we have not worked with the material before, we cannot make such predictions.

In contrast, consider "-ness" measures such as "square-ness." Despite the fact that the "square-ness" task was presented with a finite number of rectangles, we did not infer a rule for quantifying "square-ness" from these few rectangles. What we did was *assert* a rule for calculating the value of the measure we defined as "square-ness." This is in contrast to the density case, where we were dealing with real objects made from real materials and we were obliged to devise our rule from limited data. In the case of rectangles, by asserting a rule for calculating the "square-ness" measure that applies to all rectangles, there are no "*as yet untried*" *instances*. Thus, "square-ness" is not a model—it is simply a measure.

In the case of our measure of "difficulty of stopping a moving object," we have a related situation. We have defined the value of the momentum of *any* moving object in a given reference frame by saying that it can be calculated from the mass and the speed of that object. There are no "*as yet untried*" *instances*. Momentum, by itself, is not a model—it is a measure. It took an Isaac Newton to assert a model—i.e., that the time rate of change of this measure called momentum is equal to the applied force.

Getting from Measures to Models

The centrality of "*as yet untried instances*" to this discussion is by now evident. It is a manifestation of the need for models to be falsifiable (e.g., Popper, 1959). A measure whose value may always be calculated from its definition is not a model. Does this mean that measures are not germane to a discussion of models? Not at all. Measures are the constructs we need to formulate in order to build models. A model in physics, for example, may predict how a measure such as momentum will change with time. A model in chemistry may predict how a measure such as concentration of a species might change with temperature. A model in economics might predict how a measure such as GNP might change with unemployment rate. The essential point is that in order to have a model, we must be able to predict the values of a measure in instances that did not enter into the formulation of that measure. For example, in the case of density, we found our empirical data for the case of aluminum to be well described by

$$M \text{ (mass in grams)} = 2.73 \text{ grams/cc} \times V \text{ (volume in cc)}$$

The assertion that this relationship is valid beyond the data points that were used to construct this description is a model for aluminum. The assertion that for any homogeneous material the mass of an object is proportional to its volume is a generalization of this model to all homogeneous materials. The further assertion that for any homogeneous material the constant of proportionality is inversely proportional to the temperature is a still further generalization of the model.

Ideally, one would like to predict how the values of the measure one formulates might vary with time or temperature or some other parameter(s) of interest. However, we may not understand the underlying science well enough to be able to make quantitative predictions. Students, and even scientists, may, however, have a qualitative understanding of underlying mechanisms that is good enough to say that the value of a measure will increase or decrease with time or temperature. Although such semi-quantitative models have been the subject of some considerable investigation (e.g., Miller, Ogborn, Briggs, Brough, Bliss, Boohan, Brosnan, Mellar, & Sakonidis, 1993; see also Jackson, Stratford, Krajcik, & Soloway, 1995; Ogborn, 1994), there is little systematic research on how students formulate the measures whose behavior in time these models explore.

Agent-Based Models—From Local Interactions to Global Measures of Complex Behavior

Up to this point, I have discussed measures that for the most part can be thought of as macrosopic measures—measures that are characteristic of an entire system and do not vary from one part of the system to another. Needless to say, this is not always the case. Density can depend on position, as was discovered when a satellite first orbited around the moon and found that the mass of the moon was not homogeneously distributed (Grushinskii & Sagitov, 1962).

The explosive growth of computational power that has become increasingly widespread in the past few decades has enabled us to approach the problem of modeling phenomena of all sorts in a fashion that differs greatly from that which has been discussed so far. "Agent-based" models are particularly valuable in modeling systems whose properties of interest may vary from place to place. Such models are built on the assumption that phenomena can be simulated by considering a large number of agents that interact with one another in space and time via interactions that can be simply stated and whose consequences may be computed simply. Although the possibility of such models was first described in the 1940s, it was not until the widespread availability of computers that it became realistic to model phenomena in this fashion.

A person building an agent-based model posits the nature of the interactions between and among the individual agents. The model is then run and what emerges is observed and considered—sometimes it is an emergent pattern, sometimes an uncontrolled growth, sometimes the vanishing of one or more kinds of agents, sometimes an equilibrium, and sometimes an uninterpretable mess.

A typical example of a situation that lends itself well to modeling with agent-based models is that of the population of two species, one of which preys on the other. Using very simple probabilistic rules about the birth and death rates of each of the species, and the likelihood of a predator encountering prey, the evolution of the two populations can be modeled.

When this is done, one can discover that under some sets of initial conditions extraordinary behaviors emerge—e.g., predator and prey population sizes oscillating in time; other sets of initial conditions will lead to uncontrolled growth of the prey population; and so on.

The statement of the individual predator and prey rates and likelihoods of encounters are the starting point of such models. The—as yet unexamined—entities in this model are the global emergent behaviors and the measures thereof that emerge only after the individual predator and prey behaviors are allowed to develop. Some of these measures include equilibrium states, frequency, amplitude, and relative phase of the oscillatory behavior of populations, conditions for collapse of one or both species, and so forth. This is in contrast to the kind of model discussed earlier in which measures are formulated ab initio and then their time or other parametric behaviors are predicted.

SOME IMPLICATIONS FOR K–12 SCIENCE AND MATHEMATICS EDUCATION

There has been a great deal of attention paid to the central role of model building in science education (Lehrer & Schauble, 2010). While some of this work hints at the role of measures and the pedagogic power and utility of having students formulate their own measures, there seems to be no systematic attempt to explore how these measures are built up from sensory percepts, quantified, and compounded into more complex measures. By introducing the distinction between measures and models, parsing the nature of simple and composite measures, and focusing on two of the simplest possible composite measures, i.e., product measures and quotient measures, it is my hope that it will become easier to pay more explicit attention in K–12 science to the formulation of measures and their role in the formation of models. This in itself is a worthy effort that deserves attention, as is apparent from even a brief consideration of some of the Balanced Assessment "-ness" tasks described above.

Moreover, separating the task of formulating a measure from that of formulating a model that uses that measure focuses attention on the distinction between the measure construct and the model construct. This is likely to result in a better understanding on the part of teachers and students of the differences and similarities between mathematics and science. But that is the subject of another discussion!

NOTES

1. See, for example, discussion of "Evolution of Biological Complexity," http://en.wikipedia.org/wiki/Evolution_of_complexity

2. See, for example, http://web.mit.edu/murj/www/v07/v07-News/v07-world.pdf

3. It would seem that occasionally one encounters a count noun modified by a non-integer, as in the statement, "The average family in that area has 1.9 children." As we shall see later, the semantic structure of the referent situation in such cases is very different from those situations in which a counting act leads to the standard {cardinality, definition} structure (e.g., Schwartz, 1996).

4. The definition of this sort of measure is difficult for many people. Note that common thesaurus synonyms for *momentum* include *energy, force, impulse, power,* and *thrust!* All of these measures differ from momentum (and one another) and relate to one another in different ways in the building of models.

5. For those unfamiliar with the distinction between intensive and extensive quantities, see, for example, Schwartz, 1996.

6. See, for example, http://en.wikipedia.org/wiki/Gross_Domestic_Product for a simple explanation of the way in which the "Gross Domestic Product" (related to, but not identical to, Gross National Product) measure is constructed.

REFERENCES

Balanced Assessment in Mathematics. (2007). Retrieved from http://balancedassessment. concord.org

Finesilver, C. (2009). Representational strategies of students with difficulties in mathematics: Responses to a "Cartesian product" problem. *Proceedings of the British Society for Research into Learning Mathematics 29*(2), 29–34.

Fulcrum Institute. (2010). *Summer energy workshop.* Unpublished manuscript. Medford, MA: Tufts University.

Grushinskii, N. P., & Sagitov, M. U. (1962). Some considerations of the lunar gravitational field. *Soviet Astronomy, 6*(1), 113–117.

Heymans, M., & Singh, A. K. (2003). Deriving phylogenetic trees from the similarity analysis of metabolic pathways. *Bioinformatics, 19*(1), i138–i146.

Jackson, S., Stratford, S. J., Krajcik, J., & Soloway, E. (1995). *Making system dynamics modeling accessible to pre-college science students.* Paper presented at the annual meeting of the American Educational Research Association, San Francisco, CA.

Kloos, H., Fisher, A., & Van Orden, G. (2010). Situated naïve physics: Task constraints decide what children know about density. *Canadian Journal of Experimental Psychology: General, 139*(4), 625–637.

Lehrer, R., & Schauble, L. (2010). What kind of explanation is a model? In M. K. Stein & L. Kucan, (Eds.), *Instructional explanations in the disciplines* (pp. 10–22). Dordrecht, Netherlands: Springer.

Miller, R., Ogborn, J., Briggs, J., Brough, D., Bliss, J., Boohan, R., Brosnan, T., Mellar, H., & Sakonidis, B. (1993). Educational tools for computational modelling. *Computers in Education, 21*(3), 205–261.

Ogborn, J. (1994). Modeling clay for computers. *Tijdschrift voor Didacttiek der β-wetenschappen, 12*(1), 5–20.

Popper, K. (1959). *The logic of scientific discovery.* New York: Basic Books.

Schwartz, J. L. (1996). *Semantic aspects of quantity.* Retrieved from http://gseweb.harvard.edu/~faculty/schwartz/papers.htm

Schwartz, J. L. (2007). Models, simulations, and exploratory environments: A tentative taxonomy. In R. A. Lesh, E. Hamilton, & J. J. Kaput, (Eds.), *Foundations for the future in mathematics education* (pp. 161–172). Mahwah, NJ: Lawrence Erlbaum Associates.

Steptoe, A., & Feldman, P. J. (2001). Neighborhood problems as sources of chronic stress: Development of a measure of neighborhood problems, and associations with socioeconomic status and health. *Annals of Behavioral Medicine, 23*(3), 177–185.

Wong, D. W. (2005). Formulating a general spatial segregation measure. *The Professional Geographer, 57*(2), 285.

An Example of the Tasks
Used in Research

Problem 2

Knouse and Campbell (1971) carried out an experiment to analyze the effect of the partial delay of reward during behavior acquisition on how easily behavior can be extinguished. They trained some rats to obtain food after running through a straight corridor. The reward was delayed for 0, 8, 16, 24, 48 or 56 seconds in half of the acquisition trials. The extinction behavior was focused on the second part of the experiment and the animals were locked in the goal box for 15 seconds without receiving the reward.

   The results showed that the resistance to the extinction increased based on the duration of the delay. The longer the delay the animals experienced during some of the trials to acquire the behavior, the faster they ran during extinction and the longer the response lasted.

1.- Select which one of the five graphs we show you below most fits this result.
2.- Explain why you have chosen this graph.
3.- Explain the reasons why you rejected the other graphs.

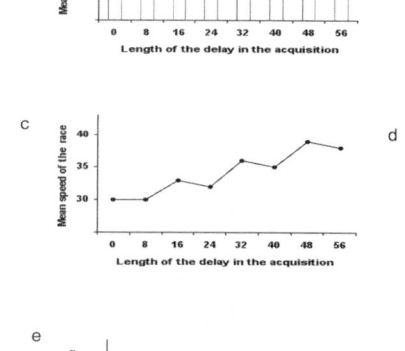
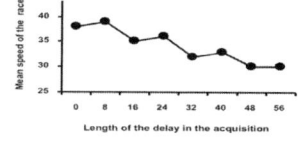
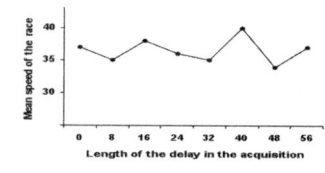
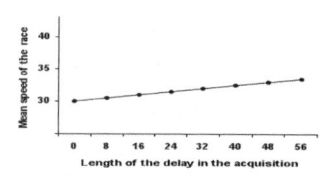

About the Editors and Contributors

Bárbara M. Brizuela is an associate professor in the Department of Education at Tufts University. Her main areas of interest are early childhood education, cognitive development, and mathematics education. Her current research focuses on children's learning of the notational number system, notational aspects of mathematics, children's learning of elementary mathematics, including early algebra, and mathematics teachers' learning and professional development.

Brian E. Gravel is a lecturer and director of elementary education at Tufts University. Coming from engineering and engineering education, his current areas of interest are elementary science education, educational technologies, and urban teacher education. His research focuses on exploring how students produce and use representations as thinking tools, educational technologies and modeling the natural world, and elementary STEM education in urban schools. Brian leverages all his work in teacher education efforts, working to create rich and meaningful STEM experiences for K–12 students.

Mary C. Caddle is a post-doctoral associate in the Poincaré Institute for Mathematics Education at Tufts University. She earned her PhD in mathematics education at Tufts University. Mary was a research assistant with the Early Algebra, Early Mathematics Project at Tufts and teaches courses for students interested in teaching middle and high school mathematics. Her dissertation research was on mapping teacher and student understanding of introductory combinatorics. Her main research interests are in analyzing student thought in algebra and combinatorics, and in the relationships between teacher and student mathematical thought.

David W. Carraher is a senior scientist at TERC and principal investigator for the Poincaré Institute for Mathematics Education, a National Science Foundation project awarded to Tufts University and TERC. He was, for 18 years, a professor, researcher, and software developer at the Federal University of Pernambuco, Brazil. His work on everyday mathematics is published in English, Portuguese, Spanish, and Japanese. At TERC since 1995, he studies early algebraic reasoning in classroom activities designed to explore the algebraic character of arithmetic as children are introduced to variables, functions, and equations. This work is described in books and articles (see www.earlyalgebra.terc.edu).

Gabrielle Cayton-Hodges is an associate research scientist in the Cognitive and Learning Sciences group at ETS. Gabrielle earned her bachelor's degree in Brain and Cognitive Sciences from MIT in 2002, and her PhD in Mathematics, Science, Technology, and Engineering Education from Tufts University in 2009 and was previously a postdoctoral research fellow at the Joan Ganz Cooney Center at Sesame Workshop. Gabrielle's main interests lie in student understandings of numerical concepts, the use of multiple representations in mathematics, and mathematical and scientific learning progressions.

Montserrat de la Cruz is a full professor of developmental psychology at the Universidad Nacional del Comahue. Her main areas of interest are children's cognitive development in the fields of drawing and writing, as well as in the development of implicit learning theories, seeking to understand the influence of schooling and sociocultural environments on these processes.

Gerald A. Goldin received his PhD in theoretical physics from Princeton University, and coordinated Mathematics Education at the University of Pennsylvania and Science Education at Northern Illinois University before joining the Rutgers University faculty in 1984. He has published approximately 200 scholarly articles in mathematics, theoretical physics, and the psychology of mathematical learning and problem solving, receiving the Humboldt Research Prize for his work in quantum physics. In mathematics education his work has focused on representations in problem solving, and the affective domain in mathematical learning and engagement.

Miguel Angel Gómez Crespo teaches physics and chemistry at Victoria Kent High School in Torrejón de Ardoz, Madrid, Spain. He received his PhD at the Universidad Autónoma de Madrid with a doctoral dissertation on Chemistry Learning and Instruction that received the Spanish National Educational Investigation and Innovation Prize in 2006. His research focuses on science learning and teaching, the use of ITC in the classroom, and Science/Technology/Society.

Tina Grotzer is an associate professor of education at Harvard University and a senior researcher at Project Zero. Her work focuses on the intersection of cognitive science, development, and science learning. She studies how people reason about causal complexity and its implications for K–12 education and the public understanding of science. She developed the *Causal Patterns in Science* curriculum and accompanying website for middle school science with support from the National Science Foundation. She is a former classroom teacher.

María Sagrario Gutiérrez Julián teaches physics and chemistry at San Juan Bautista High School in Madrid, Spain. She has specialized in the training of teacher trainers at the University of Valencia. Her research foci are teachers' instruction in science, evaluation and science teaching and learning, scientific literacy, and Science/Technology/Society.

David Hammer is a professor of education and physics at Tufts University, and co-director of the Center for Engineering Education and Outreach. His research focuses on the learning and teaching of science and engineering from K to 16.

Jason Kahn is a research associate in the Department of Psychiatry at Children's Hospital Boston and an instructor in psychiatry at Harvard Medical School. He is interested in the creation and study of tools that give young children new insight into the world around them, and continues to be amazed at the level of insight and sophisticated connections children can make when they have the opportunity to explore their world in new ways. Jason is currently working on the development and evaluation of representational tools that help children see their emotional arousal and explore ways to self-regulate.

Richard Lehrer is Frank W. Mayborn Professor of Education at Vanderbilt University. His research focuses on the design of classroom learning environments that support the growth and development of model-based reasoning in science and in mathematics. He currently investigates development of children's reasoning about space, measure, data, and chance when children participate in classrooms where instruction is guided by teacher knowledge of student reasoning. These forms of mathematical investigation support science education where, in collaboration with Leona Schauble, he works with teachers to promote children's invention and revision of models of ecosystem functioning and related building blocks of evolution.

Cristina Marín-Oller is a flutist, flute teacher, and PhD student at the Universidad Autónoma de Madrid. She is also assistant professor at the International University of La Rioja (Spain). Her main research topic is related to teaching and learning processes in music and the acquisition of its representational system, musical notation, particularly in woodwind instruments. As a member of her research group at the Universidad Autónoma de Madrid, she also studies the acquisition and use of other representational systems, like graphs and tables, among university students.

Eduardo Martí is professor of developmental and educational psychology at the University of Barcelona, Spain. He collaborated with Piaget at the Center of Genetic Epistemology (University of Geneva). His research area and publications concern cognitive development and the acquisition of external systems of representation. His last books are *Representar el Mundo Externamente* (Representing the World Externally), *Desarrollo, Cultura y Educación* (Development, Culture, and Education), and *After Piaget* (co-edited with Cintia Rodríguez).

Ricardo Nemirovsky is a professor at San Diego State University and director of the Center for Research in Mathematics and Science Education (CRMSE). He has directed educational projects in Argentina, Mexico, and the U.S. His teaching includes courses on theories of learning, geometry, and educational technology. He is currently conducting microethnographic research and developing theories on embodied cognition in mathematics learning. In addition to research papers, he has co-authored curricular units and has designed multiple devices for students' use.

Tracy Noble is a Project Leader at TERC, where she has worked since 1993. Dr. Noble's research explores the range of ways in which children learn about and express their understandings of science and mathematics, including talk, gesture, interactions with physical artifacts, and interaction with written assessments. Dr. Noble's current work focuses on how students who are learning English interact with science test items from the 5th grade MCAS, and on the linguistic features of these items that may interfere with the ability of these students to demonstrate what they know about science.

María-Puy Pérez-Echeverría has been a faculty member in the School of Psychology at the Universidad Autónoma de Madrid since 1987. Her research focuses on three interrelated subjects: external representations as learning tools (Graphs, Numbers, and Musical Scores), teachers' and students' conceptions about learning and teaching, and mathematical problem solving. She has published or co-edited three books and over 60 papers on these topics.

Merredith Portsmore, M.D., is associate director for the Tufts Center for Engineering Education and Outreach as well as a Research Assistant Professor in Education. Merredith received all four of her degrees from Tufts (BA in English, BS in mechanical engineering, MA in education, PhD in engineering education). Her research interests focus on how children create solutions to engineering design problems. Her outreach work focuses on creating resources for K–12 educators to support engineering education in the teacher education and classroom practice. She is also the founder of STOMP (stompnetwork.org) and LEGOengineering.com.

Yolanda Postigo has a PhD in psychology and is a professor at the School of Psychology at the Universidad Autónoma de Madrid, Spain. Her research focuses on the learning and use of several external representation systems (graphs, tables, images, and maps) in different domains of knowledge with students from different education levels (primary, secondary, and higher education).

Juan Ignacio Pozo is a professor at the School of Psychology of the Universidad Autónoma de Madrid, Spain, where he teaches topics related to psychology of learning. His research focuses on the processes of knowledge acquisition and conceptual change in a number of specific domains (e.g., physics, chemistry, history, geography, grammar, music). He has also studied how to change teachers' and learners' conceptions of learning as a requirement for school and curriculum transformation. He has published several books about knowledge in specific domains, conceptual change, and learning conceptions.

Ann S. Rosebery is co-director of the Chèche Konnen Center at TERC. Her research has two goals. One is to document and characterize the wide-ranging intellectual resources that children from communities historically placed at risk in our society bring to the study of science, and how these overlap with scientific ideas and practices. A second is to develop approaches to professional learning that enable teachers to (a) attune their eyes, ears, hearts, and minds to their students' powerful intellectual resources; and (b) experiment with expansive classroom pedagogies that engage their students' ideas and experiences in scientifically meaningful ways.

Leona Schauble is a cognitive developmental psychologist with research interests in the relations between everyday reasoning and schooled forms of thinking, especially scientific and mathematical reasoning. She completed her PhD in developmental and educational psychology at Columbia University in 1987. After five years at the Learning Research and Development Center at the University of Pittsburgh, she spent the next ten years of her career at the University of Wisconsin. She is currently professor of Education in the Department of Teaching and Learning at Vanderbilt University's Peabody College.

Nora Scheuer is a researcher at the National Council of Scientific and Technological Research of Argentina (CONICET), at Universidad Nacional del Comahue in Bariloche, Patagonia. Her main areas of interest are children's cognitive development in the fields of number, drawing, and writing, as well as in the development of implicit learning theories, seeking to understand the influence of schooling and sociocultural environments on these processes. Nora Scheuer earned her Doctorate in Psychology from Geneva University (studying children's appropriation of the base 10 numerical system), received her BA in educational psychology in Buenos Aires, and carried out a research apprenticeship in Italy studying young children's emerging explaining competence (Trieste University) and kindergartners' early text composition strategies (La Sapienza di Roma University).

Analúcia D. Schliemann was a professor of psychology at the Federal University of Pernambuco, in Brazil, and a Professor of Education at Tufts University. As Professor Emerita at Tufts, she is currently a principal investigator in the Poincaré Institute for Mathematics Education, a National Science Foundation sponsored project. Her publications on everyday mathematics and on children's early algebraic reasoning appeared in multiple periodicals and in books such as *Street Mathematics and School Mathematics* (with T. Nunes and D. Carraher, Cambridge University Press, 1993) and *Bringing Out the Algebraic Character of Arithmetic* (with D. Carraher and B. Brizuela, Lawrence Erlbaum Associates, 2007).

Judah L. Schwartz is a theoretical physicist who has studied people's understanding of mathematics and science for more than 35 years. Until recently he was senior lecturer in the Department of Education and research professor in the Department of Physics at Tufts University. Before that he was professor of Engineering Science and Education at MIT and professor of Education at Harvard University. His research focuses on the design of interactive computer software to augment intuition about abstract constructs in science and mathematics.

Michael A. Smith is a doctoral student at the Center for Research in Mathematics and Science Education, a joint research program between San Diego State University and the University of California at San Diego. His research focuses on combining phenomenology, embodied cognition, neurology, and research psychology to explore the nature of mathematical thinking and difficulty.

Beth Warren is co-director of the Chèche Konnen Center at TERC. Her research centers on four themes: (a) documentation of the wide-ranging, intellectually powerful sense-making repertoires of students from historically non-dominant communities, (b) analysis of generative intersections between students' sense-making repertoires and those used routinely in academic disciplines, (c) exploration of classroom practices that build on heterogeneity as a first principle of design for expansive learning, and (d) development of a practice-based approach to professional learning integrating subject matter, learning, pedagogy, and structural inequalities on the same plane of professional inquiry.

Kristen Wendell has a PhD and is an assistant professor of elementary science education in the Department of Curriculum and Instruction and the Center of Science and Mathematics in Context at the University of Massachusetts Boston. Kristen's research focuses on the affordances and constraints of integrating engineering design into children's science, reading, and writing experiences in the elementary grades. She also studies the perceptions, knowledge, and skills that pre-service elementary teachers have related to engineering design. Her background includes aerospace engineering, science curriculum design, and teacher professional development in science and engineering.

Christopher G. Wright, a postdoctoral fellow with the Chèche Konnen Center (TERC), explores issues of equity and social justice in science and engineering education. His current research examines the intellectual, representational, and linguistic resources that African American boys bring to the study of science, specifically, the extent to which these resources reflect the practices utilized within professional STEM communities.

Index